# 中非可再生能源合作报告 2023

水电水利规划设计总院　非洲联盟发展署　编

中国水利水电出版社
www.waterpub.com.cn
· 北京 ·

**图书在版编目（CIP）数据**

中非可再生能源合作报告. 2023 : 汉文、英文 / 水电水利规划设计总院，非洲联盟发展署编. -- 北京 : 中国水利水电出版社，2025. 1. -- ISBN 978-7-5226-2795-3

Ⅰ. F426.2；F446.2

中国国家版本馆CIP数据核字第2024V218W5号

审图号：GS京（2024）2280号

责任编辑：郝英　张晓

| 书　名 | **中非可再生能源合作报告 2023**<br>ZHONG-FEI KEZAISHENG NENGYUAN HEZUO BAOGAO 2023 |
|---|---|
| 作　者 | 水电水利规划设计总院<br>非洲联盟发展署　编 |
| 出版发行 | 中国水利水电出版社<br>（北京市海淀区玉渊潭南路1号D座　100038）<br>网址：www.waterpub.com.cn<br>E-mail：sales@mwr.gov.cn<br>电话：（010）68545888（营销中心） |
| 经　售 | 北京科水图书销售有限公司<br>电话：（010）68545874、63202643<br>全国各地新华书店和相关出版物销售网点 |
| 排　版 | 中国水利水电出版社微机排版中心 |
| 印　刷 | 北京科信印刷有限公司 |
| 规　格 | 210mm×285mm　16开本　10.25印张　292千字 |
| 版　次 | 2025年1月第1版　2025年1月第1次印刷 |
| 定　价 | **198.00**元 |

编委会

在全球能源转型的大背景下，可再生能源发展已成为推动经济增长和实现可持续发展的重要动力。非洲是全球可再生能源资源最为丰富的大陆之一，水能、风能和太阳能资源在全球范围内占据重要地位。 非洲联盟（以下简称“非盟”）一直致力于推动非洲大陆的能源转型和可持续发展，发展可再生能源根植于非盟《2063 年议程》优先项目，是非洲各国领导人通过非盟采纳的长期愿景，并与非洲大陆对繁荣和包容性增长的期许高度一致。《非洲基础设施发展计划》（Programme for Infrastructure Development in Africa，PIDA）也将能源作为重点发展领域之一。 2021 年，非盟峰会上通过了 PIDA 第二期优先行动计划（2021—2030 年）的 71 个优先项目，覆盖能源、水利等领域，包括冈比亚河流域组织的水力发电项目等，涉及非盟 40 余个国家，在推动地区一体化、环境友好型等方面具有典型代表意义和规模效应。

可再生能源的开发利用对于满足非洲日益增长的电力需求、改善民生和推动工业化进程具有重要意义。 非洲国家领导人在多个国际场合强调了在可再生能源领域的潜力和决心，呼吁国际社会支持非洲的可再生能源发展。 近年来，非洲各国加快了可再生能源发展的步伐，水电、风电和太阳能发电装机容量显著增加，2016—2023 年，非洲可再生能源装机容量（不含抽水蓄能）占能源装机总量的比例从 20.2%增长到 24.6%。 2023 年，非洲气候峰会通过《非洲领导人关于气候变化的内罗毕宣言及行动呼吁》（简称《内罗毕宣言》），呼吁国际社会协助非洲提升可再生能源发电能力，以期从 2022 年的 56GW 提升至 2030 年的 300GW。

中国始终是非洲国家可再生能源发展道路上的主要合作伙伴，为非洲基础设施和可再生能源领域发展作出巨大贡献。 早在 2016 年，非盟基础设施和能源事务委员易卜拉欣博士就指出：“非洲约 30%的能源项目是由中国企业执行的。”大多数非洲国家正处于转型升级发展的关键时期，中国在可再生能源领域的先进经验正是非洲国家在发展中需要学习和借鉴的。 非洲各国期待与中国在可再生能源领域开展更多务实合作，实现经济、社会和环境的可持续发展，为非洲的绿色未来奠定坚实基础。 未来，非盟将继续推动非洲可再生能源发展，期待与国际社会，特别是中国等合作伙伴，通过政策交流、战略对接、机制建设和项目合作等方式，共同推动非洲能源转型，朝着非盟《2063 年议程》描绘的美好愿景加速前进，全力建设和平、团结、繁荣、自强的新非洲。

拉赫曼塔拉·奥斯曼

非洲联盟驻华代表处常驻代表

# 前言

过去十年间，非洲经济展现出较强的发展韧性和活力，预计将持续成为全球经济增长最快的地区之一。随着非洲联盟（以下简称“非盟”）《2063 年议程》稳步推进，非洲大陆自由贸易区正式实施，次区域组织相互协作不断加强，非洲正在成为具有全球影响的重要一极。在追求经济增长的同时，这片大陆不仅需要将气候行动融入更广泛的社会和经济发展活动中，还需要采取有效的适应措施，以应对气候变化的不利影响。通过利用新技术、推广可再生能源、提高旧能源系统的效率以及改变管理实践和消费者行为等措施，可以有效减缓气候变化的影响。

非洲致力于构建有韧性的低碳经济体系。非洲拥有全球约 12%的水能、32%的风能和 40%的太阳能资源，发展可再生能源具有天然优势。非洲的电力普及率相对较低，可再生能源的开发利用具有重要的战略意义，不仅能够满足电力需求、改善民生，还能够推动经济发展、实现可持续发展。

当前，应对气候变化和实现可持续发展已成为全球共识，绿色低碳转型正在世界范围内不断加速。中国是全球可再生能源发展的重要力量，可再生能源开发利用规模稳居世界第一。截至 2023 年年底，中国可再生能源总装机容量达 1517GW，约占全球可再生能源总装机容量的 40%，为全球能源绿色转型提供强大支撑。中国与非洲国家不断加强可再生能源合作，在保障能源供应安全的同时，有效推动非洲绿色低碳转型与可持续发展，取得一系列合作成果，为推动中非战略合作、实现可持续发展提供了新动力，展现了优势互补、互利共赢的合作局面。

为有效推动中非能源合作，2021 年 10 月，中国国家能源局与非盟委员会建立中国-非盟能源伙伴关系（本段简称“伙伴关系”），受中国国家能源局委托，水电水利规划设计总院（以下简称“水电总院”）配合牵头开展伙伴关系筹建和运行有关工作，在伙伴关系框架下推动中非能源领域相关政策、战略规划和项目信息等方面的全面交流，助力非洲国家政府部门和能源企业有关人员提升管理能力和专业技术水平，共同实现能源可持续发展的多重目标。2024 年恰逢第九届中非合作论坛会议在中国举办，水电总院与非洲联盟发展署联合编写《中非可再生能源合作报告》，系统梳理中非可再生能源合作的丰硕成果，提出推动中非可再生能源合作高质量发展建议和愿景，衷心希望为中非双方开展可再生能源领域合作提供有益的参考。

非洲可再生能源资源丰富，开发程度较低，发展潜力巨大。近年来非洲经济快速发展，能源需求不断加大，非洲各国将发展可再生能源作为促进能源转型的重要途径。联合国非洲经济委员会（UNECA）也指出，可再生能源是满足非洲大陆快速增长的电力需求的可行替代能源。中非在可再生能源领域合作基础良好，中国企业在非洲建设了数百个水电、风电、太阳能发电等可再生能源项目，为非洲国家缓解能源短缺、实现绿色发展提供助力。本报告梳理了非洲可再生能源发展和中非可再生能源合作情况，总结中非双方可再生能源领域合作机遇和挑战，提出合作建议，展望合作愿景，希望为双方可再生能源合作和可持续发展提供指导和实践参考。

### 1. 非洲经济持续增长，用能需求不断释放

近年来非洲经济展现出较强的发展韧性和活力，2019—2023 年，面对新冠疫情、地缘冲突和全球金融紧缩等多重挑战，非洲实际 GDP 平均增长率为 2.1%，与全球平均增速持平。联合国经济及社会理事会（UNESC）预测，2024 年非洲经济将增长 3.5%，增速在全球范围内仅次于亚洲。经济的稳步增长将促使非洲不断释放用能需求。根据非洲大陆电力系统总体规划（CMP）预测，到 2040 年，非洲用电量将达到 3842TWh，最大负荷为 634GW。随着非洲各国不断推动可再生能源开发，据国际可再生能源署（IRENA）预测，到 2030 年，非洲可以通过使用可再生能源来满足其近 1/4 的能源需求。

### 2. 非洲各国加快可再生能源开发，有效推进能源转型

目前非洲能源消费和生产仍以化石能源为主，人均用电量相对较低；同时，非洲拥有丰富的可再生能源资源，各国正在加快能源转型，推动可再生能源的开发利用。2023 年非洲发电总装机容量为 252.8GW，其中，可再生能源（不含抽水蓄能）装机容量合计为 62.1GW，约占总装机容量的 24.6%。近年来，非洲可再生能源（不含抽水蓄能）装机增速明显高于化石能源发电装机增速，2019—2023 年，非洲可再生能源（不含抽水蓄能）装机容量增长率达 23.2%，大大超过化石能源装机容量增速（6.4%）。根据 CMP 预测，到 2040 年，非洲电源装机总规模将达到 1200GW，可再生能源装机容量将达到 750GW，占总装机容量的 62.5%，非洲能源结构将发生较大转变，向更加清洁和可持续的能源系统过渡。

### 3. 中非可再生能源合作机制建设日臻完善，项目合作持续深化

自新中国成立至今，70 多年来中非双方风雨同舟、携手前行，中国致力于不断巩固中非政治互信，深化各领域务实合作，为非洲和平与发展提供力所能及的帮助，中国对非合作一直走在国际对非合作的前列。中国不断深化“一带一路”能源合作伙伴关系和全球可再生能源合作伙伴关系，通过联合国、国际可再生能源署、二十国集团、金砖国家等合作平台，积极参与全球能源治理。在《非

洲可再生能源倡议》的指导下，在“一带一路”倡议和中非合作论坛的引领下，在多双边合作机制的推动下，中非可再生能源领域技术不断创新，项目合作开花结果，合作模式从对外援助、工程承包为主，向推动投建营一体化方向发展。过去几十年，中国企业积极响应“走出去”战略，在非洲建设了数百个可再生能源和绿色发展项目，项目类型涵盖水电、风电、太阳能、生物质能等领域，取得了一系列合作成果，不仅注重经济效益，更注重社会效益和环境效益，赢得了当地政府和民众的高度评价，助力非洲走绿色可持续发展之路。

#### 4. 把握机遇互利共赢，应对挑战共谋发展

在全球大力发展可再生能源的背景下，中非可再生能源合作面临新的发展机遇。非洲亟须发展可再生能源来满足日益增长的用能需求，中国在可再生能源领域技术持续发展，中非双方互补优势不断扩大，可再生能源合作正当其时。然而，国际形势不断变化、资金短缺、基础设施不足、人才资源缺乏等问题仍是合作需面临的挑战。建议中国与非洲加强政策沟通，深化机制合作，促进技术交流与人才培养，推动绿色金融模式创新，提升项目经济可持续性，打造合作示范项目，加强同非洲国家相互支持和友好合作，共同应对风险挑战。

#### 5. 加强务实合作，共同谱写中非可再生能源合作新篇章

中非合作论坛第八届部长级会议上，中国宣布将与非洲国家密切配合，共同实施包括绿色发展工程在内的“九项工程”，为非洲国家经济社会发展创造了新机遇，且进一步释放了中非可再生能源合作的潜力，助力实现中非可再生能源发展愿景。为进一步推动中非可再生能源合作高质量发展，双方应加强机制建设，继续发挥中非合作论坛的战略引领作用，加强政策交流和战略对接，促进知识和经验的分享，将技术和相关产业本土化，以促进非洲国家经济的上下游联动。此外，绿色金融模式的创新也是推动合作的关键，中国可以与非洲国家共同探索适合当地市场的绿色金融产品，为可再生能源项目提供资金支持。此外，中非双方应加强人才培养和技术支持，通过技术培训等措施，支持非洲国家提升可再生能源领域的专业能力，提升项目的经济可持续性，构建知识共享和技术创新的桥梁。通过这些措施，中非可再生能源合作将更加深入和广泛，不仅促进了双方的共同发展，也为全球能源转型和应对气候变化作出了积极贡献。中非双方将继续在现代化道路上携手向前，不断提升可再生能源合作的水平和层次，共筑高水平中非命运共同体。

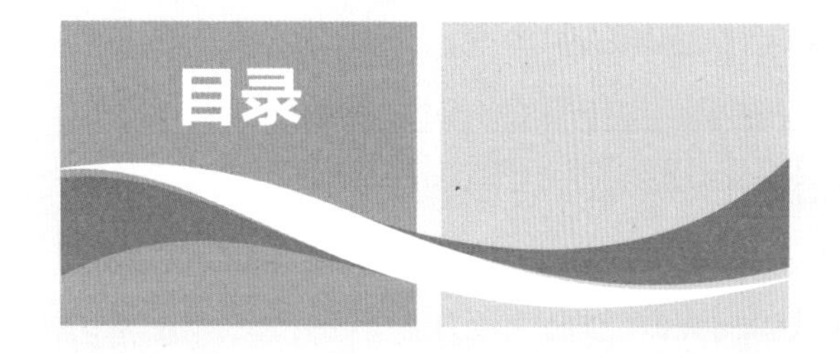
目录

# 1 非洲概况

非洲位于东半球的西南部，横跨赤道南北，东濒印度洋，西临大西洋，北隔地中海与欧洲相望，东北与亚洲红海相隔，以苏伊士运河为陆上分界，总面积约为 3037 万 $km^2$，约占地球陆地总面积的 1/5，是世界第二大洲（见图 1.1）。非洲大陆形状北宽南窄，平均海拔 750m，是世界上平均海拔最高的大陆之一，海岸线总长 30500km，气候以高温、少雨、干燥为特点，大部分地区年平均气温在 20℃以上，部分地区终年炎热，被誉为“热带大陆”。

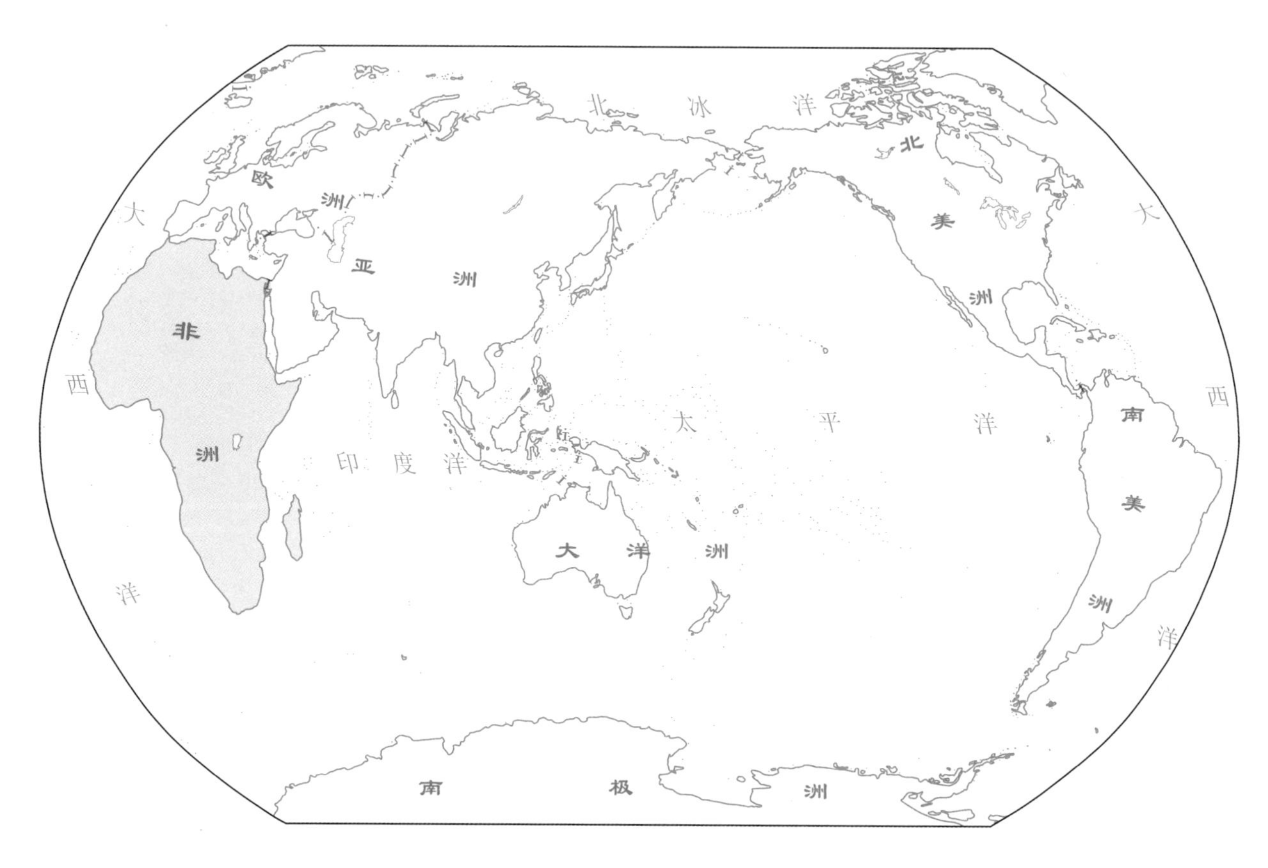


图 1.1 非洲地理位置图

非洲是主权国家分布最多的大洲。为了便于分析，联合国按照地理位置，并结合自然条件及社会经济等特点，将非洲分为五大区域，分别是北部非洲、西部非洲、中部非洲、东部非洲和南部非洲（见图 1.2）。

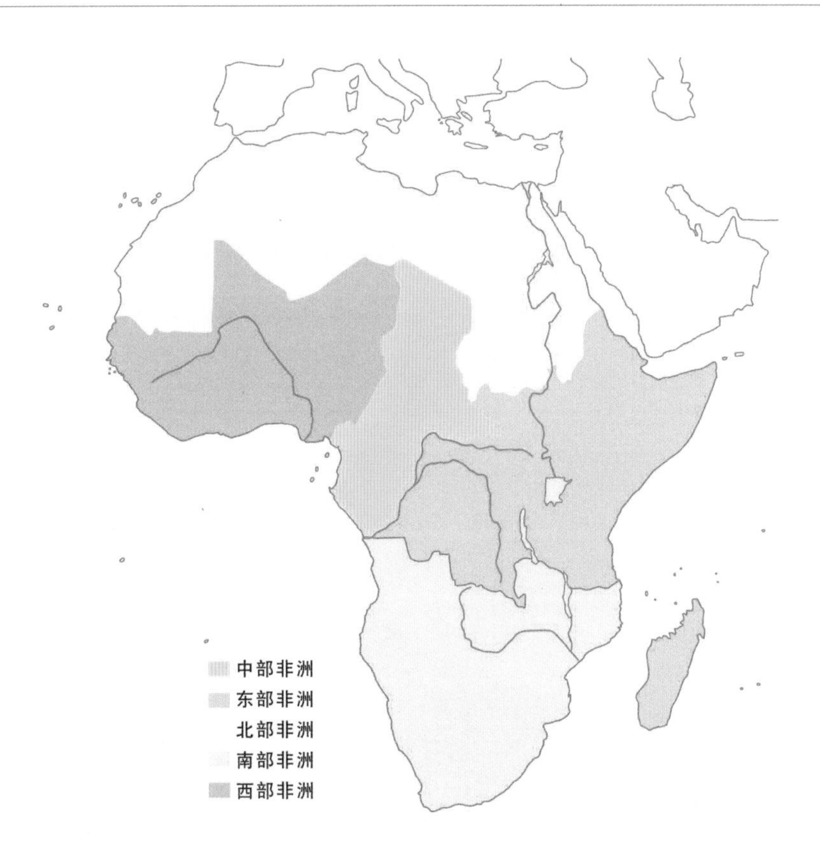


图 1.2 非洲地理区域划分图（数据来源：UNDESA 地理方案）

## 1.1 非洲人口红利将持续释放

截至 2023 年，非洲大陆人口超过 14 亿人，约占世界总人口的 1/6，居世界第二位，仅次于亚洲。非洲是人口增长率最高的地区，根据联合国经济和社会事务部（UNDESA）人口统计司预测，到 2030 年非洲人口将达到 17 亿人，到 2040 年非洲人口将突破 20 亿人（见图 1.3）。

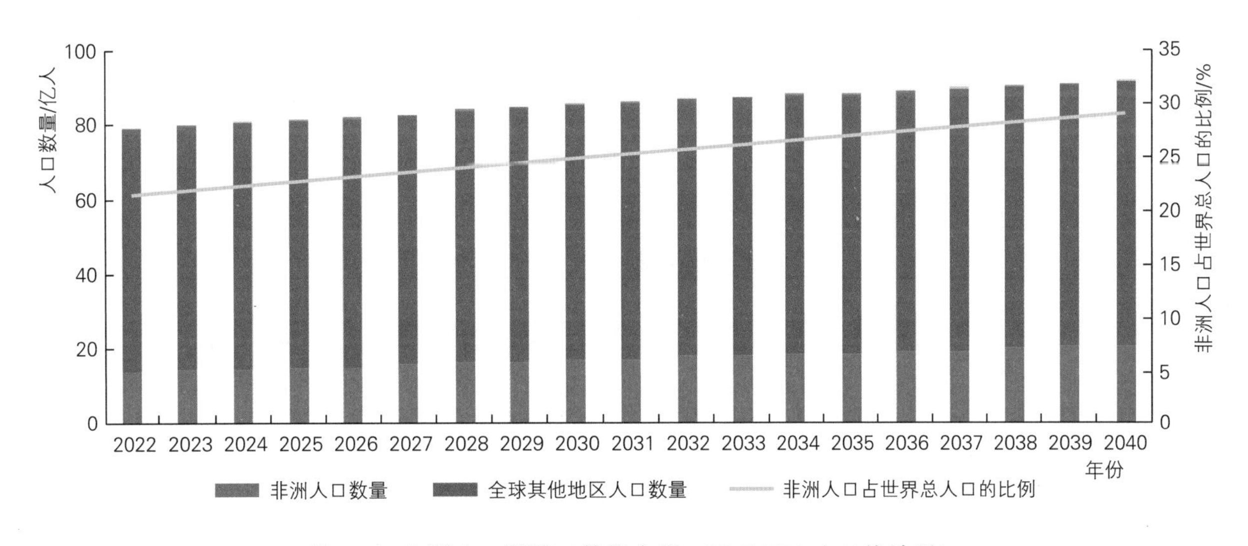


图 1.3 非洲人口预测（数据来源：UNDESA 人口统计司）

根据联合国《世界人口展望 2024》报告，截至 2023 年，非洲劳动力（15～64 岁）人口达到 8.44 亿人，约占非洲总人口的一半以上，其中青年（15～24 岁）人口约为 2.89 亿人，占非洲总人口的 1/5（见图 1.4）。预计到 2050 年，非洲工作年龄（20～64 岁）人口将增至 16 亿人，占全球工作年龄人口的近 1/4，而青年人口将增加至 4.27 亿人，意味着全球每三名年轻人中就有一名来自非洲。非洲庞大的年轻人口预示着在未来数十年里非洲将拥有丰富的劳动力资源和巨大的消费市场潜力。这一人口结构的优势将持续释放人口红利，为非洲的经济增长提供源源不断的动力和活力。

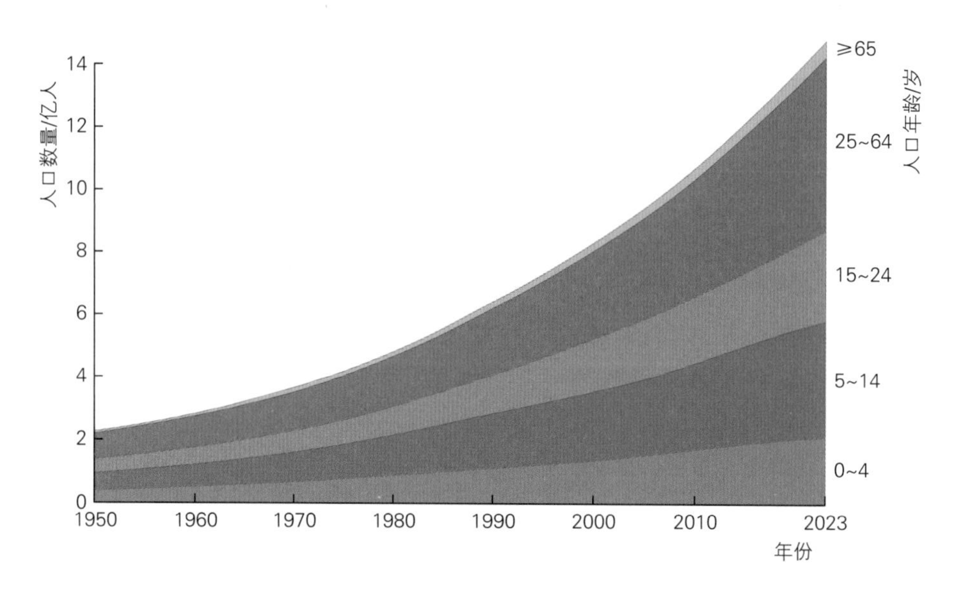


图 1.4　非洲人口结构情况（数据来源：联合国《世界人口展望 2024》）

## 1.2　非洲经济预计将长期快速增长

2023 年非洲 GDP 总量达到 2.75 万亿美元，人均 GDP 为 1987.5 美元，约为全球人均 GDP 的 15%。在新冠疫情、地缘冲突、极端气候频发、全球金融紧缩等多重挑战下，非洲经济保持韧性，据统计，2019—2023 年非洲实际 GDP 平均增速为 2.1%，与全球平均增速相当（见图 1.5）。

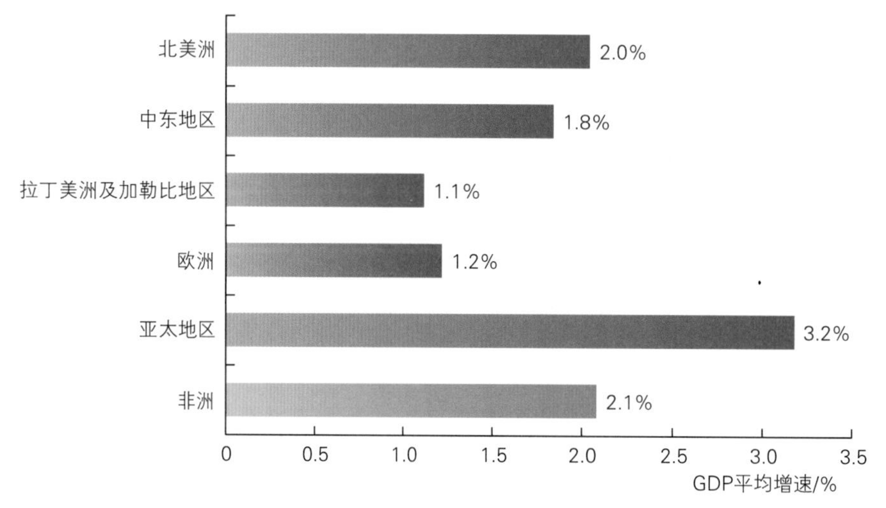


图 1.5　2019—2023 年全球主要地区 GDP 平均增速
（数据来源：经济学人智库）

近年来，由于全球地缘局势日益紧张，各国对自身能源、矿产供应安全担忧加剧，加大对关键资源竞争。非洲能源、矿产资源储量丰富，相关领域出口和外商投资加速增长，部分非洲国家加大能源、交通等基础设施建设投资，2021—2023 年，投资和出口对非洲经济增长贡献持续上升。

产业结构上，服务业是非洲经济的重要支柱，占非洲 GDP 一半以上（见图 1.6），主要得益于近年来旅游业复苏以及金融业和信息技术的快速发展。采掘业也发展迅速，但本土加工能力有限，多为原材料出口，提升本土加工能力和增加产品附加值是非洲经济发展的关键方向之一。制造业基础较为薄弱，占比仅略高于 10%。非洲需要进一步推动经济多元化，特别是加强制造业和其他产业的发展，以实现经济的均衡增长，可以利用其在服务业和自然资源方面的优势，与其他国家进行国际合作，引进技术和资本，促进产业的升级和发展。

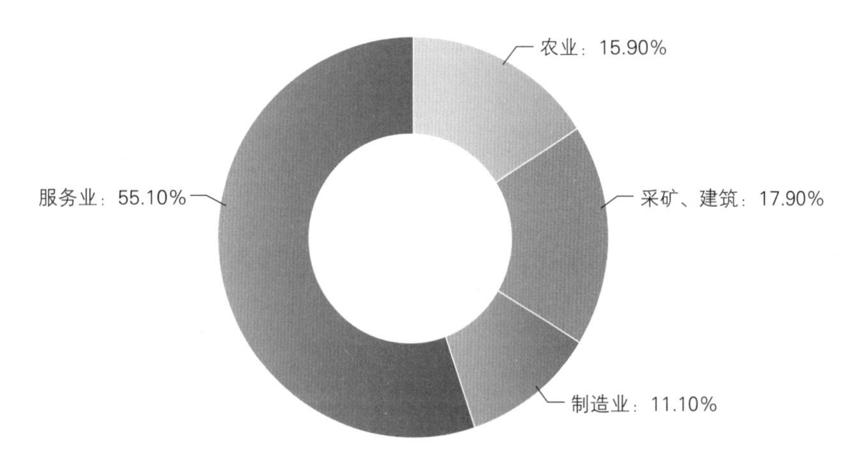


图 1.6　非洲产业结构（数据来源：经济学人智库）

随着非洲大陆自贸协定一、二阶段谈判基本完成，全球最大的自由贸易区正在成形。到 2030 年，非洲大陆自贸区将消除 90%以上货物的关税，区域内贸易占比有望较快提升。根据世界银行预计，到 2035 年，全面贸易自由化将使非洲实际收入提高 7%，总出口量增加 29%，区域内贸易占比从当前的 13%上升到 20%（见图 1.7），经济产出预计增加 2120 亿美元。区域一体化将降低非洲内部贸易壁垒，促进非洲经济资源的自由流动和优化配置，有利于各国参与市场竞争和分工合作，充分发挥自身比较优势，长期来看有望助推非洲的工业化进程。

根据经济学人智库预测，2024—2025 年，非洲经济有望反弹，预计平均增长率将达到 3.4%。这一积极趋势主要得益于全球对非洲丰富自然资源的需求增加，以及货币政策从紧缩逐步转向宽松

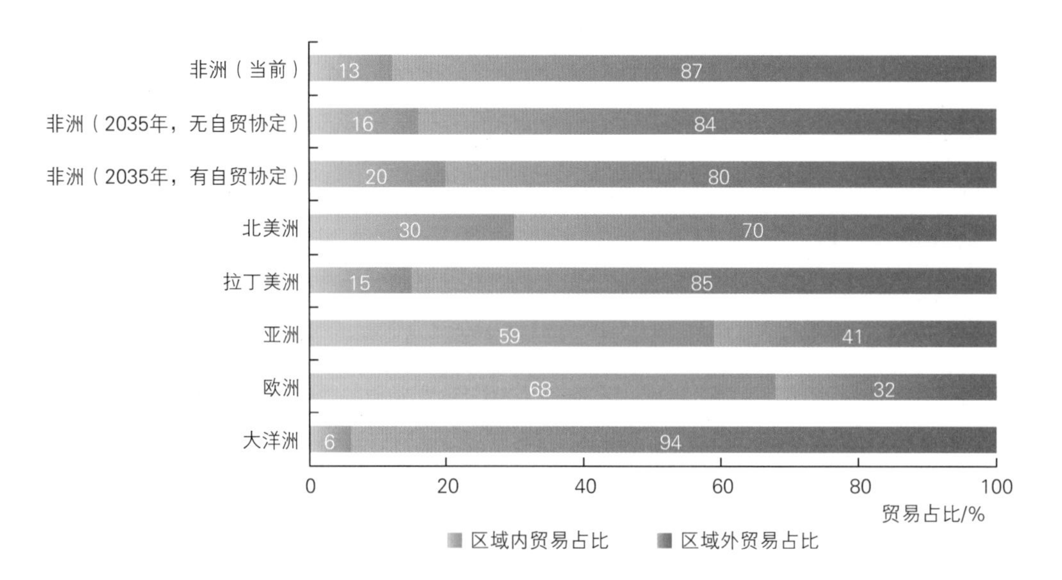


图 1.7　非洲区域内、外贸易占比

［数据来源：联合国贸易和发展会议（UNCTAD）、世界银行集团（WBG）］

的宏观环境。 展望未来，非洲人口增长势头强劲，预示着其经济潜力将持续释放。 2024—2040 年，非洲的年均 GDP 增速预计将达到 3.8%，较全球平均增速（2.5%）高出 1.3 个百分点（见图 1.8），非洲将成为全球增长最快的地区。

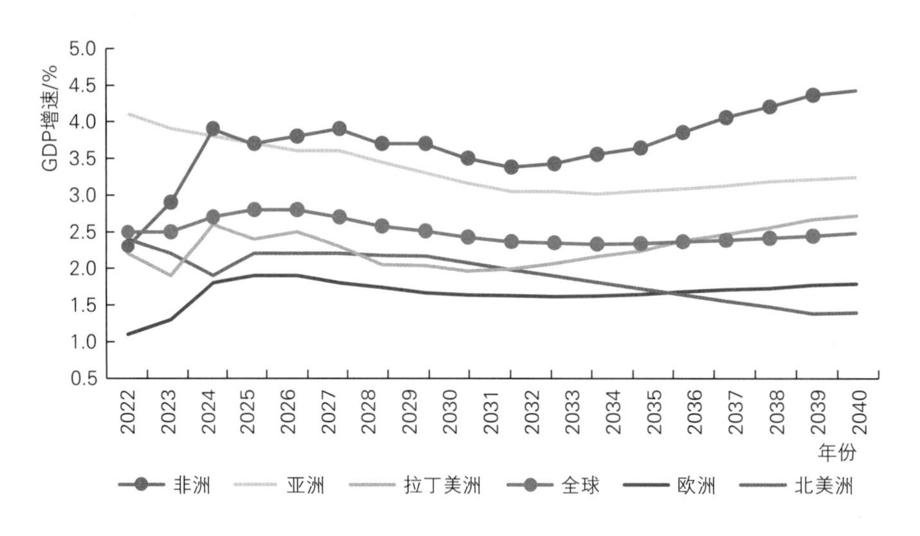


图 1.8　全球各地区实际 GDP 增速预测（数据来源：经济学人智库）

## 1.3　能源生产与消费水平有待提升

非洲大陆在能源领域表现出积极发展态势。 随着经济和人口增长、工业化进程加快以及对能源需求的增加，2011—2022 年，非洲的一次能源生产与消费量整体上呈现出稳步上升的趋势。 同时，非洲具有较强的适应性和恢复力，尽管 2020 年新冠疫情对全球经济造成了冲击，导致非洲一次能源生产和消费量暂时下降，但很快就恢复了增长。 然而值得注意的是，非洲的能源生产和消费量仍低于世界平均水平。2022 年，非洲一次能源生产量为 35.87EJ，仅占全球一次能源生产总量的 5.9%；一次能源消费量为 20.26EJ，仅占全球一次能源消费总量的 3.4%；人均一次能源消费量 13.9GJ，约为世界平均水平的 18%（见图 1.9）。

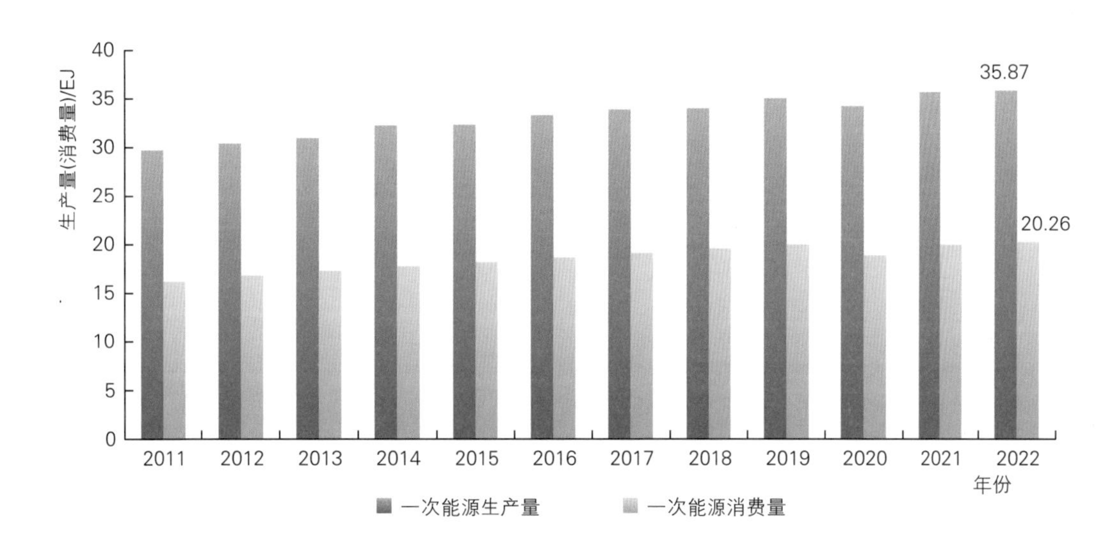


图 1.9 2011—2022 年非洲一次能源生产量和消费量
[数据来源：国际能源署（IEA）]

非洲的能源结构以化石能源和传统生物质为主。石油、天然气、煤炭、水电以及少部分核能构成了非洲能源的主要部分。从能源消费端来看，2022 年石油、天然气和煤炭消费比重分别为 41.41%、28.87%和 19.60%，可再生能源消费比重相对较低，仅占 9.67%（见图 1.10）。从能源生产端来看，2023 年其他可再生能源（主要是太阳能和风能）生产比重仅占 1.6%，相当于 0.1 亿桶油当量。此外，水电占比为 2.5%，石油占比为 45.8%，天然气占比为 32.3%，煤炭占比为 17.3%。根据 IEA 研究显示，到 2028 年，天然气占比将超过石油，2033 年其他可再生能源占比将超过水电，2042 年将超过煤炭（见图 1.11）。非洲在可再生能源领域的发展潜力巨大。

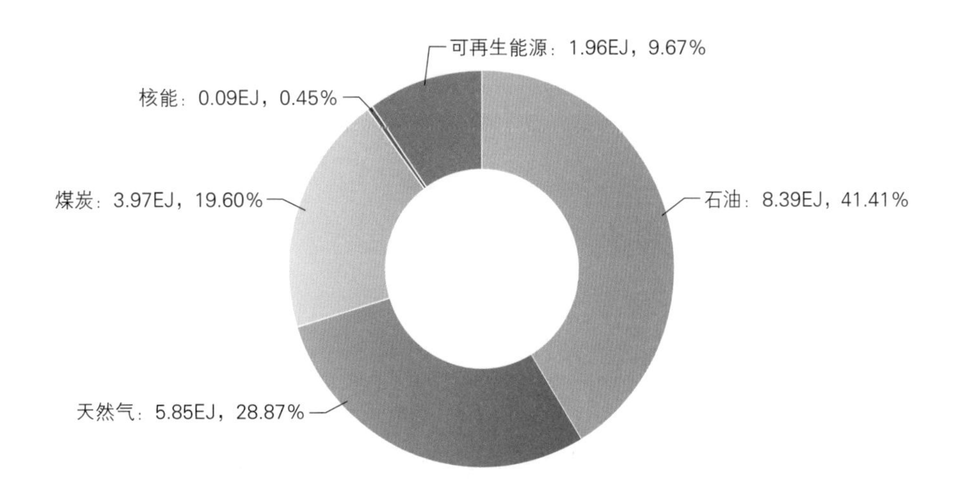


图 1.10 2022 年非洲各能源消费量及结构（数据来源：IRENA）

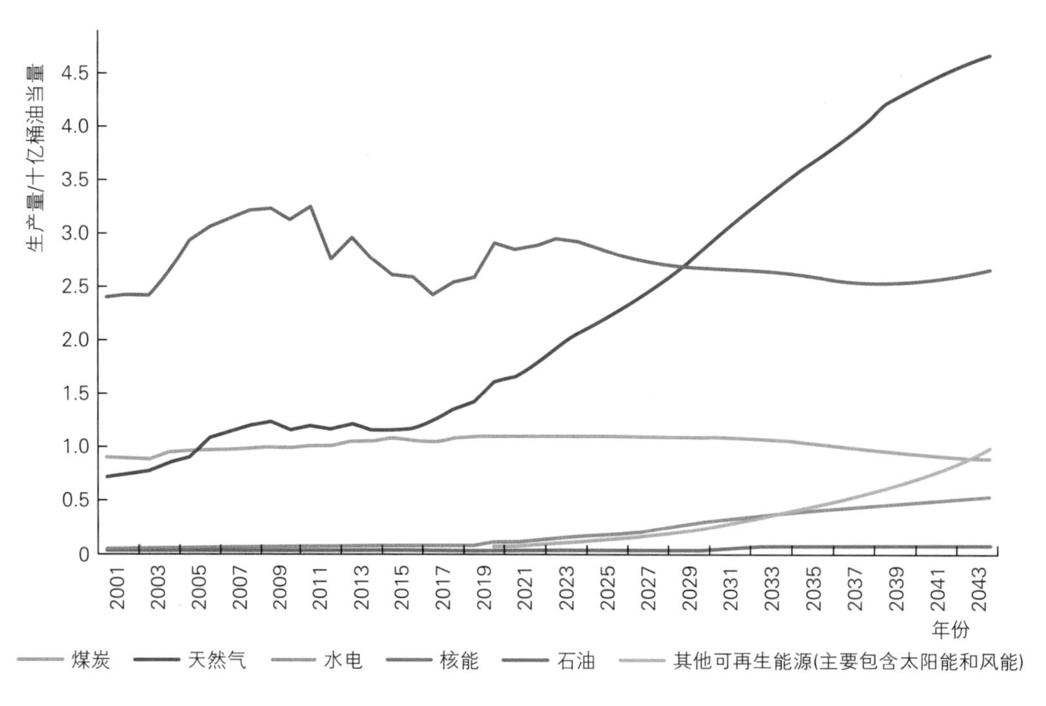


图 1.11 2001—2043 年非洲各能源生产量变化趋势预测（数据来源：IEA）

## 1.4 非洲国家能源进出口情况存在较大差异

总体来看，非洲是一个能源净出口大陆，大部分煤炭、未精炼石油和天然气在非洲当地生产，出口到欧洲和亚洲市场。根据 IEA 相关研究，到 2042 年，非洲能源出口额和能源进口额将趋于平衡（见图 1.12）。到 2043 年，阿尔及利亚、尼日利亚、安哥拉和莫桑比克将成为非洲前四大能源出口国，预计出口额分别为 702 亿美元、320 亿美元、242 亿美元和 145 亿美元。南苏丹（59 亿美元）、刚果共和国（57 亿美元）、乍得（39 亿美元）和加蓬（27 亿美元）也将成为能源出口额较大的非洲国家。

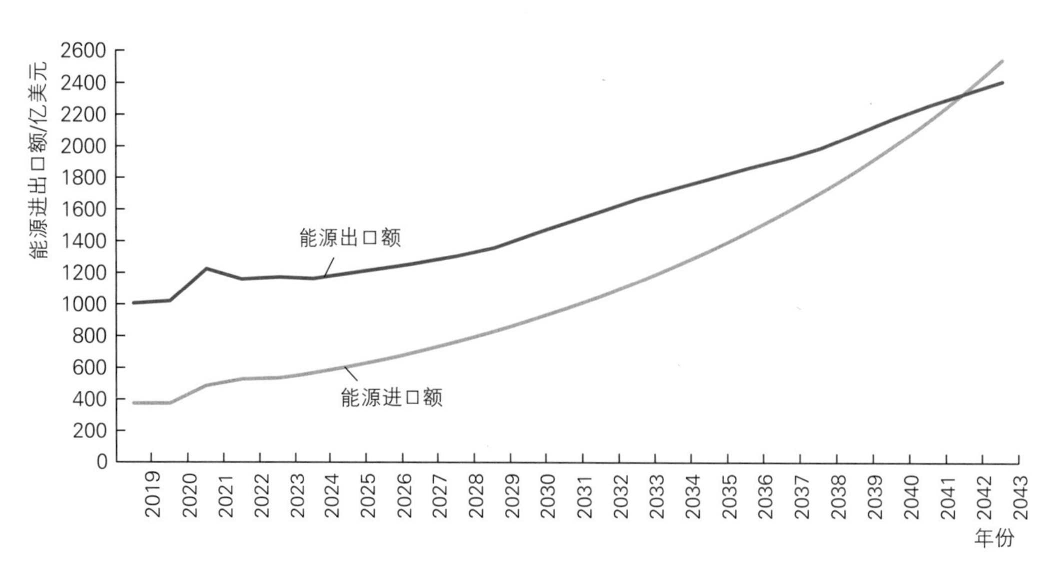


图 1.12 2019—2043 年非洲能源进出口额变化趋势（数据来源：IEA）

但非洲煤炭和油气资源分布不均，集中在少数国家，其他多数国家依旧高度依赖能源进口，可能面临能源安全威胁。目前，一些非洲国家能源进口量占其需求的50%以上，例如塞内加尔（占其需求的95%）、摩洛哥、贝宁、毛里塔尼亚、毛里求斯、厄立特里亚、突尼斯、乌干达、马里和多哥（占其需求的59%）。预计到2043年，布基纳法索、科特迪瓦、赞比亚、埃塞俄比亚、埃及、尼日尔、马达加斯加、苏丹、卢旺达、利比亚、喀麦隆和吉布提等国能源进口量也将面临同样的问题。随着可再生能源成本的逐渐降低，非洲大部分国家有望通过加大可再生能源利用力，缓解对能源进口依赖，实现能源的自给自足。

## 1.5 发展可再生能源是促进非洲可持续发展的重要途径

2023年非洲发电总装机容量约为252.8GW，其中，可再生能源装机容量62.1GW，约占非洲发电总装机容量的24.61%，包括水电（不含抽水蓄能）装机容量37.1GW、风电装机容量8.7GW、太阳能装机容量13.5GW、生物质能装机容量1.9GW、地热能装机容量1.0GW；核能装机容量1.9GW，约占非洲发电总装机容量的0.75%；化石能源装机容量188.7GW，约占非洲发电总装机容量的74.64%（见图1.13）。

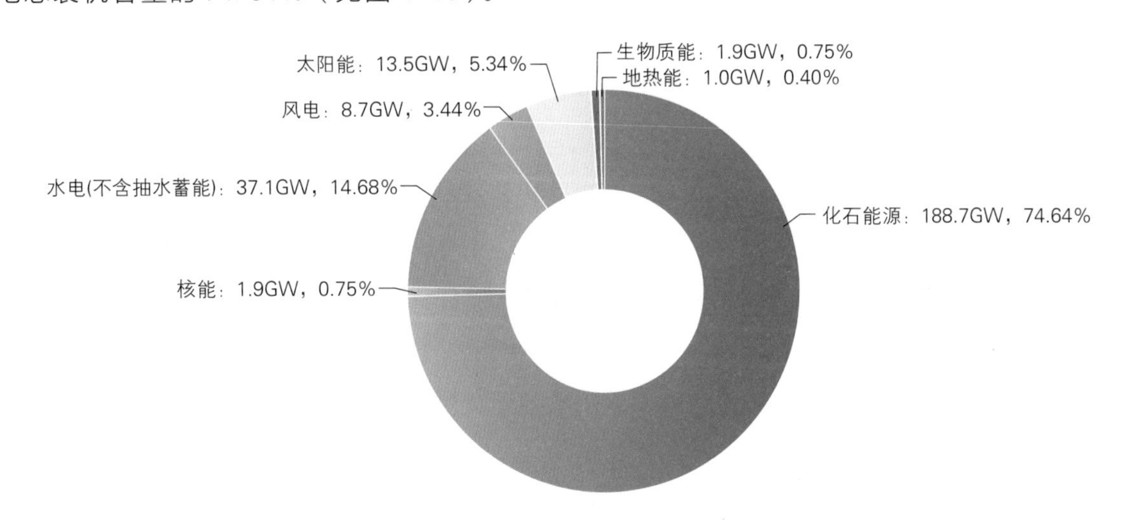


图1.13 2023年非洲各类电源发电装机容量及占比（数据来源：IRENA）

根据相关研究，2023年非洲发电总量为865TWh，其中，可再生能源发电量202TWh，约占非洲发电总量的23.35%，包括水电发电量159TWh、风电发电量14TWh、太阳能发电量21TWh、生物质能发电量3TWh、地热能发电量5TWh；核能发电量13TWh，约占非洲发电总量的1.50%；化石能源发电量650TWh，约占非洲发电总量的75.15%（见图1.14）。

目前，非洲有5个电力池组织，分别为东部非洲电力池（EAPP）、北部非洲电力池（COMELEC）、中部非洲电力池（CAPP）、西部非洲电力池（WAPP）和南部非洲电力池（SAPP），涵盖了非洲49个国家。另外，毛里求斯、马达加斯加、佛得角等岛屿国家有独立电网运行。当前，尽管各电力池成员国之间以及电力池之间均建成（或已规划）电力线路进行电能交换，但是非洲无电、少电等问题依旧严峻，电力发展不平衡、电网接入不充分等问题依然突出。

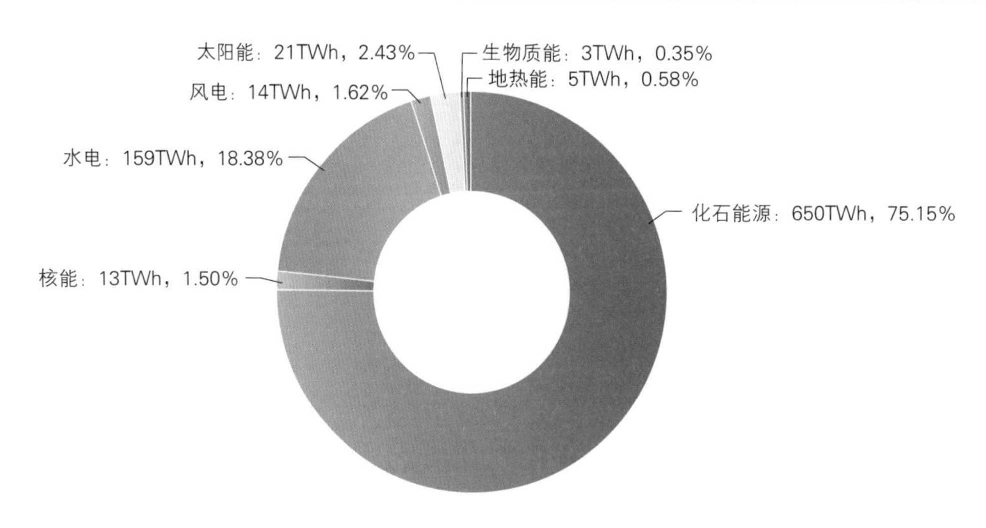


图 1.14　2023 年非洲各类电源发电量及占比（数据来源：水电总院）

据 IEA 研究显示，2023 年非洲人均用电量约 530kWh，仅为世界平均水平的 1/5，而撒哈拉以南非洲地区人均用电量约为 190kWh（其中不包括南非）。根据非洲开发银行（AfDB）2022 年统计，非洲通电率仅 40%，超过 6.4 亿人口生活在无电地区，主要分布在撒哈拉以南的非洲地区，约 2/3 的非洲国家电力可及率低于 50%（见图 1.15）。非洲电力供应不足已成为制约其可持续发展的关键瓶颈，严重制约非洲经济社会发展。

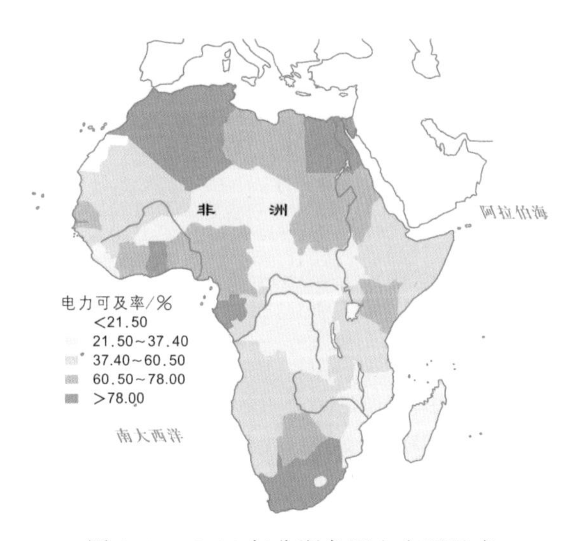


图 1.15　2022 年非洲各国电力可及率（数据来源：世界银行）

非洲正在大力发展可再生能源，把可再生能源发电作为其未来电力系统的重要支柱之一。非洲联盟（以下简称“非盟”）于 2018 年提出构建非洲大陆统一电力系统（AfSEM），旨在提升电力可及性，支持可持续发展。2023 年 9 月，非盟批准了非洲大陆电力系统总体规划（CMP），该规划开展了三种情景下的电力发展预测，分别以非洲在 2030 年（高速增长）、2035 年（中速增长）和 2040 年（低速增长）全面实现电力可及为目标（见图 1.16）。根据中增速情景方案，预计到 2040 年，非洲用电量将达 3842TWh，最大负荷 634GW，总装机容量 1200GW，其中可再生能源发电装机容量将从 2023 年的 62.1GW 增长至 750GW，可再生能源发电装机容量占非洲电力总装机容量的比例也将由 2023 年的 24.6% 增长至 62.5%，光伏发电和风电装机容量预计将分别增长至 262GW 和 342GW（见图 1.17）。

为了解决非洲电力供应不足的问题，推动经济和社会可持续发展，非洲正在依托其能源资源特点，大力发展可再生能源，并加强非洲各国电网基础设施建设，通过改造、升级和新建输配电网，扩大电网覆盖范围、提升供电效率；同时加快推动洲内及跨洲电力互联，实现非洲电力资源优化配置。

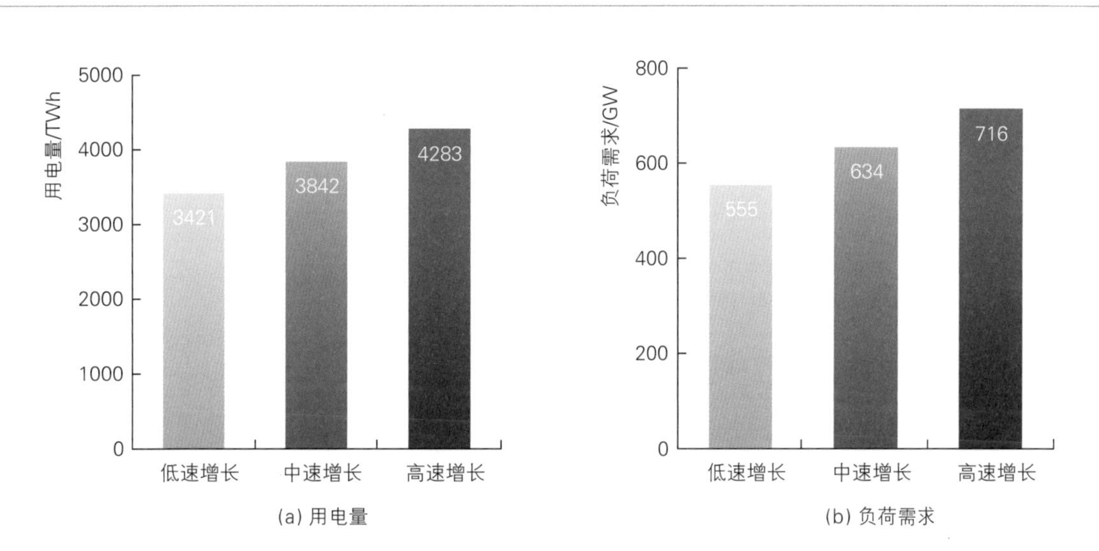


图 1.16　CMP 中三种不同情景下预测结果（数据来源：CMP）

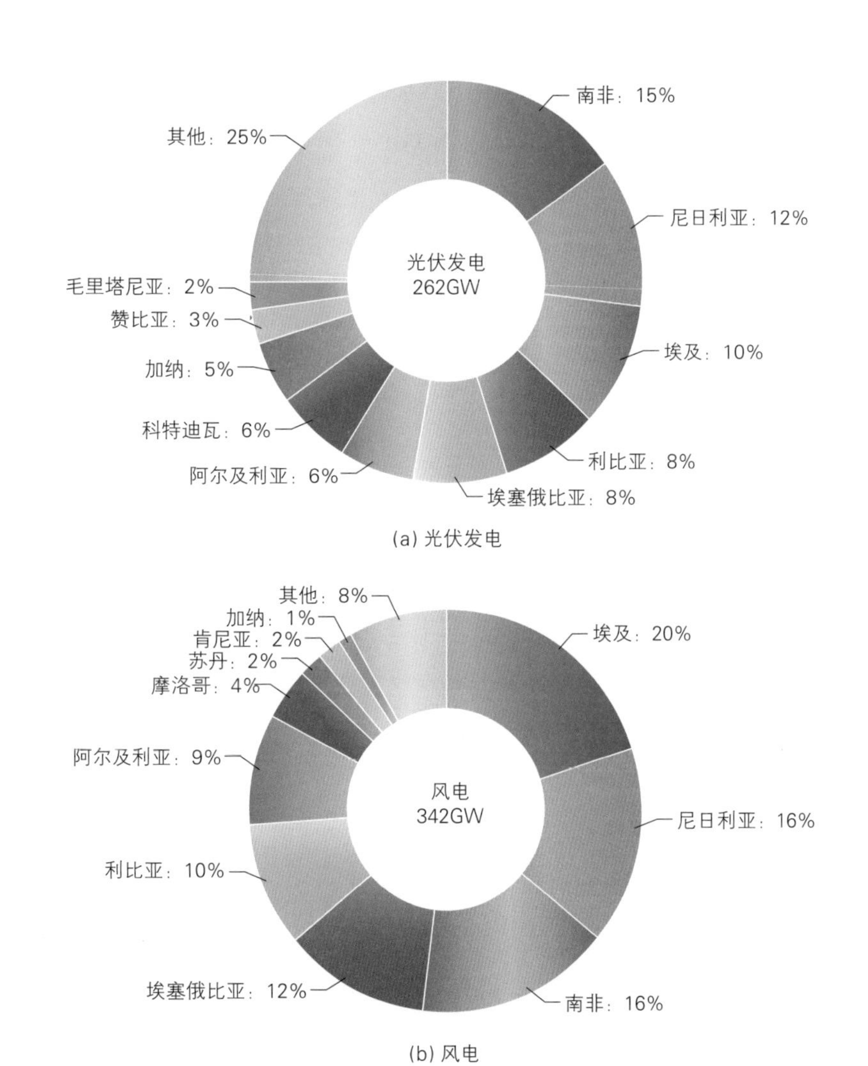


图 1.17　预测到 2040 年非洲光伏发电和风电装机分布情况
（数据来源：CMP）

发展可再生能源对于促进非洲地区可持续发展具有重要意义。非洲大陆拥有丰富的可再生能源资源，这些资源的开发和利用有助于提高非洲能源自给率，减少对化石燃料的依赖，同时减少温室气体排放，应对气候变化。此外，可再生能源项目通常具有较小的环境影响，有助于保护非洲的生态系统和生物多样性。通过投资可再生能源，非洲国家可以创造就业机会，促进当地经济发展，并提高能源的普遍可及性，特别是在偏远地区。

## 1.6 非洲城市发展与电力消费多维分析

### 1.6.1 非洲城市化趋势

当前非洲城市化进程正在以前所未有的速度发展，据相关预测，到 2050 年，60% 的非洲人口将居住在城市地区。这一现象有两个主要推动因素，一是人口由农村向城市迁移，二是城市人口自然增长，这对基础设施特别是电力消费产生了深远的影响。根据《非洲城市化动态 2020 报告》，未来几十年城市人口将翻倍，这将对经济发展、环境可持续性和社会保障产生广泛影响。

目前，45% 的非洲人口居住在城市中心。城市是电力消费的主要来源，城市化进程加快将对电力供应系统施加巨大压力。特别是在未来几年，城市用电量预计将大幅增加，需要在基础设施、能源发电和配电网络方面加强投资，才能满足不断增长的城市人口用电需求。

### 1.6.2 城市化与用电量

预计到 2040 年，非洲的电力需求将增长 5 倍，从 2022 年的 700TWh 增长到 2368TWh。城市将在这一增长中发挥重要作用，随着电力获取的改善和经济条件的加强，非洲城市人均用电量（2023 年为 530kWh）预计将持续上升。尽管如此，非洲城市地区人均电力消费量仍显著落后于全球平均水平 3000kWh。

城市电力消费受到多种因素的影响，包括人口增长、经济活动和工业化。在拉各斯、金沙萨和内罗毕等非洲城市，满足日益增长的电力需求不仅需要电网扩张，还需要整合可再生能源和智能基础设施投资。这些努力对于构建现代化城市电网和解决许多非洲城市存在的频繁停电情况至关重要。

### 1.6.3 电力基础设施的投资需求

为了满足预期的电力消费增长需求，非洲每年需要在电力基础设施上投资 4000 亿～7000 亿美元。这一投资对于扩大电网、推动电网现代化、整合可再生能源和确保城市能够满足日益增长的电力需求至关重要。到 2040 年，城市电力消费预计将增长超过 250%，这一挑战将需要政府和发展伙伴协调努力并调动必要的财政资源。

此外，非洲的贫民窟扩张和电力分布不均问题进一步加剧基础设施挑战。约翰内斯堡和开罗等城市由于电网负荷过重而频繁停电。上述例子强调全面城市规划的必要性，包括可持续发电和配电的相关规定。

### 1.6.4 未来十年的挑战

非洲城市面临的最紧迫的挑战之一是电力接入问题。尽管非洲在电气化方面取得了显著进展，但仍有超过6亿非洲人口缺乏可靠的电力供应。随着越来越多的人口迁往城市，城市化将加剧这些挑战。因此，非洲国家必须致力于推进雄心勃勃的电网现代化计划，特别是在不适宜进行电网扩张的非正规定居点，促进可再生能源的使用，并采用微电网等创新解决方案。

此外，可持续发展仍是重中之重。非洲严重依赖化石燃料发电，但全球能源转型为转向更可持续的电力来源提供了机会。对可再生能源的投资不仅为非洲日益增长的电力需求提供了解决方案，而且与全球应对气候变化和促进可持续发展的努力相一致。

### 1.6.5 发展前景

非洲城市发展和电力消费未来前景如何，取决于非洲大陆能否具备解决其基础设施差距、推动电网现代化，并将可再生能源纳入能源结构中的能力。非洲城市是经济增长的关键驱动力，确保电力的可靠供应对其持续发展至关重要。政府、国际伙伴和私营部门须共同努力，确保进行必要的投资。通过解决电力接入、电网现代化和可持续性方面的挑战，非洲可以利用其城市转型的潜力，为后代打造一个包容和可持续发展的未来。

## 1.7 非洲的电气化和数字化：释放社会经济潜力的途径

随着城市化进程的加快、不断增长的人口和数字化转型的承诺为社会经济发展提供了新的途径，非洲正处于发展的关键时期。然而，缺乏可靠的电力供应使非洲大陆数字化和经济增长潜力长期受限。数字化及数字经济的蓬勃发展，与可靠、可负担且可持续的电力供应密不可分。随着非洲大陆寻求融入全球数字经济，电气化仍然是至关重要的先决条件。

### 1.7.1 电力和数字化基础

电力供应是现代经济的基础，然而非洲仍有超过6亿人缺乏可靠的电力供应，特别是在农村和城郊地区。电力供应不足严重阻碍非洲的数字化进程。根据美国国际开发署（2024—2034年）的数字政策，建设开放、包容和安全的数字生态系统对社会经济发展至关重要，但没有稳定充足的能源供应，这些数字生态系统将无法正常运行。

到2025年，非洲数字经济有望为非洲大陆的经济贡献1800亿美元，解决电力挑战变得更加紧迫。数据中心、云基础设施、移动网络和电子商务平台均耗能较大，需要稳定的电网才能有效运行。此外，数字经济的核心——人工智能（AI）技术的整合依赖于持续不断的电力供应。例如，如果能够合理利用AI，预计到2035年它将使非洲国家的GDP增长率翻一番。然而，实现这一目标需要持续对电气化进行投资。

### 1.7.2 电气化是数字经济的催化剂

非洲目前的人均电力消费量为 530kWh，远低于全球平均水平 3000kWh。 这一差距突显了非洲大陆数字化进程所面临的能源挑战。 数字服务、数据驱动型产业和其他形式的数字经济活动在很大程度上依赖广泛、可负担得起和可靠的电力供应。 没有可靠的电力供应，数据中心、移动网络和云计算平台等数字基础设施的扩张将仍然受到阻碍。 此外，随着非洲经济越来越依赖 AI，对电力的需求度只会越来越高。

例如，非洲联盟的大陆人工智能战略强调，AI 将成为农业、医疗保健、教育和公共服务交付等领域发展的关键驱动力。 然而这种潜力只有在具备必要的基础设施的情况下才能得以实现，其中包括可靠的电力供应。 AI 系统，特别是那些处理大型数据集或运行机器学习算法的系统，是非常耗能的。 电力不足意味着即使 AI 技术可用，其应用也将受到限制。

### 1.7.3 可再生能源的投资

非洲对可持续和可靠电力的需求为投资可再生能源创造了机遇。 非洲联盟的人工智能战略强调了可再生能源在确保非洲的数字化转型与全球可持续性目标保持一致方面的重要性。 非洲大陆 60% 以上的新电力投资必须来自可再生能源，这不仅有助于满足日益增长的能源需求，也有助于支持更广泛的环境可持续性目标。

太阳能和微电网为农村电气化提供了有价值的解决方案，进而支持在服务不足的地区扩大数字和移动服务。 这些分散的能源解决方案为弥合电力差距提供了一种实用的方法，使更多社区能够参与数字经济。 对离网和小型电网解决方案的投资可以帮助缓解非洲的电力挑战，特别是在扩展传统电网成本过高的偏远地区。

### 1.7.4 电气化和数字化面临的挑战

尽管数字技术在推动非洲社会经济转型方面具有巨大潜力，但仍面临几方面挑战。 首先，电力和数字服务的可负担性问题仍是主要障碍。 许多非洲家庭和小型企业无法负担持续获取电力，更不用说依赖它的数字服务了。 正如《非洲人工智能报告》所指出的，任何旨在促进数字增长的战略都必须解决基础设施负担能力的更广泛问题。 其次，存在分布不均的挑战。 非洲国家的电气化率存在显著差异，甚至在同一个国家内部也是如此，从而导致数字鸿沟的形成。 这种数字鸿沟在农村地区尤为突出，缺乏电力的问题因缺乏互联网基础设施而进一步加剧。 为了解决这个问题，政策制定者应优先投资于能源和数字基础设施，以确保没有人在数字经济中掉队。

### 1.7.5 未来愿景：数字化发展和电力接入

弥合电力鸿沟不仅仅是为提供电力，还要为数字创新与经济增长创造有利的环境。 非洲精通技术的年轻人有能力发挥数字技术的优势，但如果没有电力，他们的能力仍将无法得到发挥。 政府、国际发展机构和私营部门必须共同努力，确保非洲的电气化目标与数字化愿景相一致。

对可再生能源和创新电网解决方案的投资，必须与扩大互联网接入、提高数字素养以及培养可以利用数字机遇的创业生态系统相结合。例如，《非洲大陆人工智能战略》展望未来，人工智能驱动的农业、健康和教育创新将可能改变非洲社会，但前提是必须具备一定的基础设施条件，尤其是电力供应。

非洲的数字化道路与解决非洲大陆的电力挑战密不可分。可靠和可持续的能源基础设施对于为数字经济提供动力、支持数据驱动型产业和确保像 AI 这样的技术能够被充分利用以促进社会经济转型至关重要。如果不解决目前的电力赤字，特别是在农村和不发达地区，非洲就有在数字经济中落后的风险。在可再生能源、电网扩张和偏远地区电气化方面的战略投资不仅将弥合电力差距，还将在数字时代释放非洲的潜力。随着非洲大陆开始其数字化转型，电气化必须继续成为优先事项，为建设具有韧性和包容性数字经济奠定基础。

# 2

# 非洲可再生能源资源与发展概况

## 2.1 非洲可再生能源资源丰富

### 2.1.1 水能资源

非洲水能资源丰富，主要河流有世界第一长河尼罗河、世界第二大水系刚果河（扎伊尔河）以及尼日尔河、赞比西河等（见图 2.1）。非洲河流水能资源理论蕴藏量为 4 万亿 kWh，约占全球的 10%；水力资源技术可开发量约为 1.75 万亿 kWh，约占全球的 12%。截至 2022 年，非洲水能开发利用率仅为 11%，有巨大的开发潜力和开发空间。

图 2.1 非洲主要河流分布图

## 尼罗河

尼罗河位于非洲东北部，发源于布隆迪高原，自南向北注入地中海，是世界上最长的河流，全长6670km，流域面积约287万$km^2$，多年平均径流量为840亿$m^3$。尼罗河的两条主要支流为白尼罗河和青尼罗河，白尼罗河全长3700km，发源于非洲中部的大湖地区，流经乌干达、南苏丹和苏丹；青尼罗河全长1600km，发源于埃塞俄比亚高地，流经埃塞俄比亚、苏丹。

## 刚果河

刚果河也称为扎伊尔河，位于非洲的中西部地区，发源于赞比亚境内，全长约4640km，是非洲中西部最长的河流，也是非洲第二长河。刚果河流经赞比亚、刚果民主共和国、刚果共和国和安哥拉，最后流入大西洋，流域面积为370万$km^2$，年径流量为1.3万$m^3$，水能理论蕴藏量为3.9亿kW。

## 尼日尔河

尼日尔河位于赤道以北的西非地区，发源于几内亚境内的富塔贾隆高原的山地中，海拔大约为900m，河流全长约4200km，是非洲第三大河流。尼日尔河在西非画出了一个倒U形，流域面积约为209万$km^2$，流经几内亚、马里、尼日尔、贝宁和尼日利亚等国家，是西非地区的“母亲河”。

## 赞比西河

赞比西河位于非洲东南部地区，发源于赞比亚西北部边境海拔1300m处的山地，与刚果河的源头十分接近，河流全长约2660km，是非洲第四大河流，也是赤道以南非洲地区的第一大河。赞比西河的流域面积约为135万$km^2$，干流流经安哥拉、赞比亚、纳米比亚、博茨瓦纳、津巴布韦和莫桑比克等6个国家。

### 2.1.2 风能资源

非洲风能资源总量约占全球风能资源总量的 32%，总体分布不均，主要集中在撒哈拉沙漠及其北部地区、南部沿海和中东部沿海地区。其中，索马里、毛里塔尼亚、乍得、埃及等国家的部分地区 100m 高度平均风速在 8m/s 以上（见图 2.2），撒哈拉沙漠及其北部地区、索马里半岛沿岸及非洲南部沿岸地区 100m 高度风功率密度可达 $400W/m^2$ 以上（见图 2.3）。根据国际金融公司（IFC）测算，非洲风能潜力为 18000TWh/年，约为非洲大陆当前用电需求的 250 倍。非洲大陆风电技术可开发总装机容量约为 33641GW，其中，北部非洲风电技术可开发装机容量约为 18822GW，南部非洲约为 891GW，东部非洲约为 2133GW，西部非洲约为 9144GW，中部非洲约为 2651GW（见表 2.1）。

**表 2.1　非洲各区域风能技术可开发装机容量（数据来源：IFC）**

| 区　域 | 风电技术可开发装机容量/GW | 区　域 | 风电技术可开发装机容量/GW |
|---|---|---|---|
| 北部非洲 | 18822 | 东部非洲 | 2133 |
| 西部非洲 | 9144 | 南部非洲 | 891 |
| 中部非洲 | 2651 | 五区总和 | 33641 |

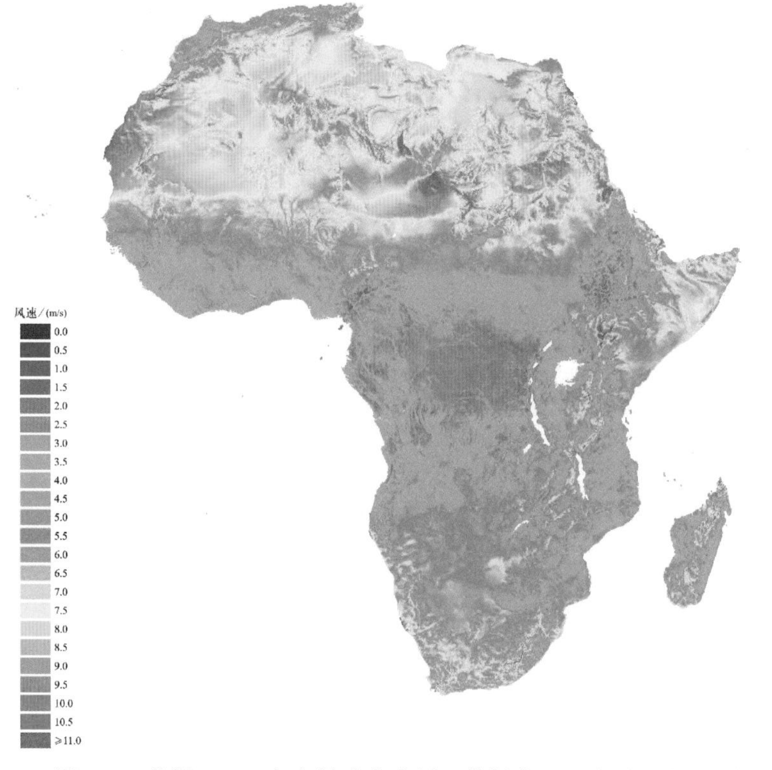


图 2.2　非洲 100m 高度风速分布图（数据来源：全球风能图谱）

### 2.1.3 太阳能资源

非洲太阳能资源分布广泛，其中光伏资源分布尤为丰富，太阳能光热和光伏的理论可开发量分别约为 470PWh/年和 660PWh/年，可开发规模约占全球太阳能资源总量的 40%。非洲 3/4 的土地可接受太阳垂直照射，大部分地区年均日照时间超过 2500h，特别是非洲北部、东部和南部的大多数国家，年平均水平面总辐照量超过 $8000MJ/m^2$（见图 2.4）。

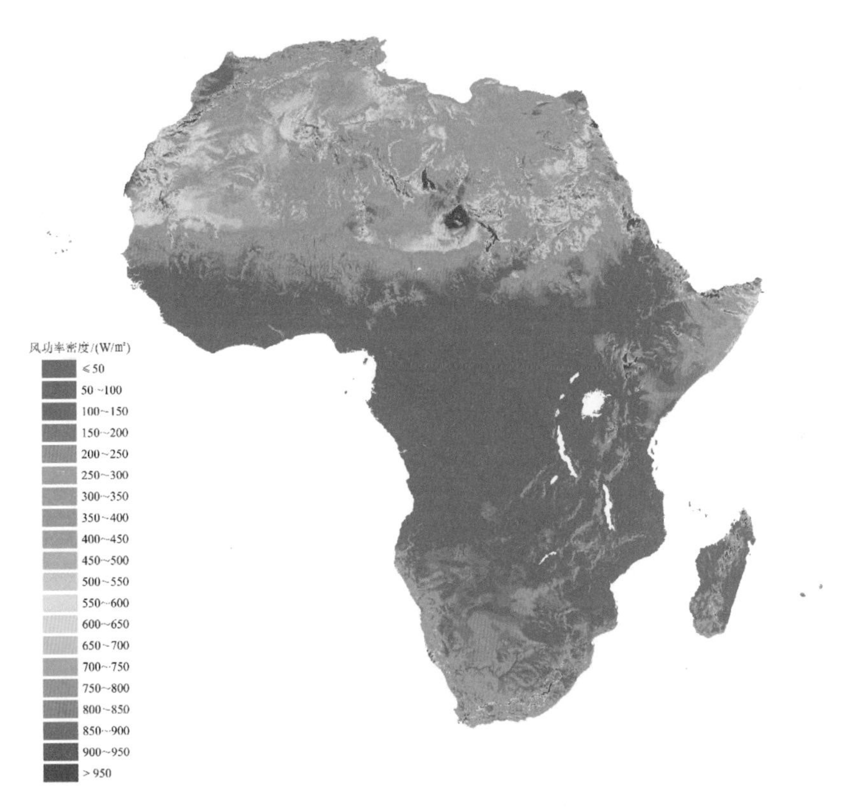


图 2.3 非洲 100m 高度风功率密度分布图（数据来源：全球风能图谱）

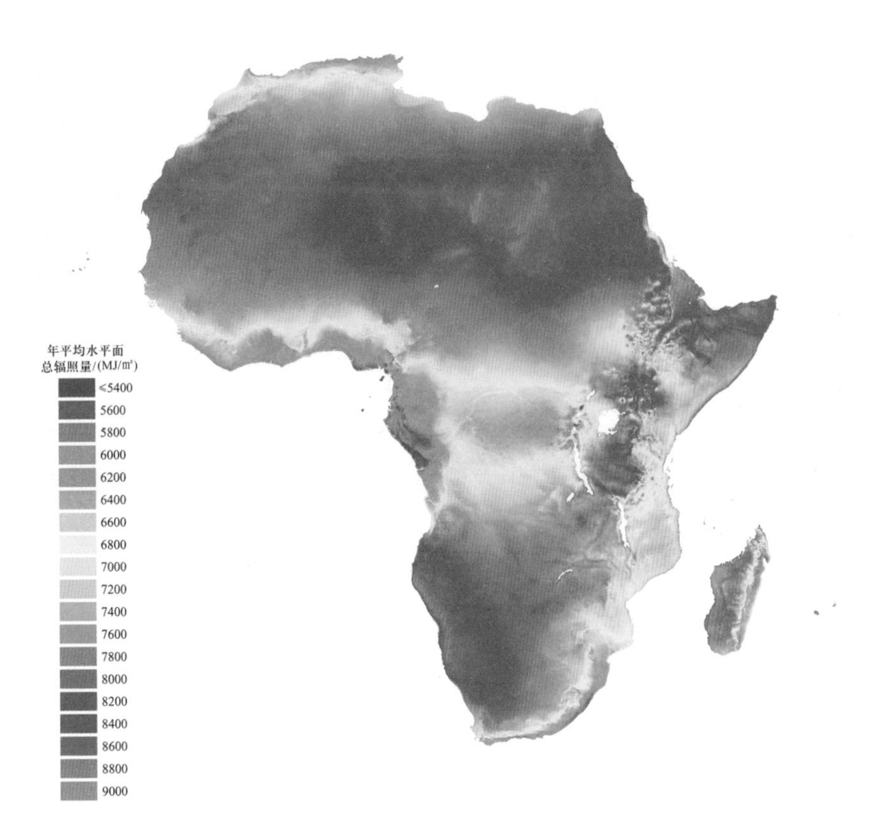


图 2.4 非洲太阳能资源分布图（数据来源：太阳能资源评估系统）

### 2.1.4 生物质能资源

非洲拥有丰富的生物质能资源，近 90%分布在撒哈拉以南地区。生物质能是非洲大陆使用最为广泛的能源，在非洲地区能源结构中占比高达 55%，但大部分都以低效的传统方法用于烹饪，并未有效地转化为电能。沼气发电是生物质能发电主要的方式之一。非洲有着高温、少雨、干燥的气候特点，具有得天独厚的气候优势，非常适合沼气发酵。据悉，东非和北非的沼气生产潜力为 $1.9\times10^6m^3/h$ 和 $1.3\times10^6m^3/h$，西非和南非生产潜力分别为 $0.4\times10^6m^3/h$ 和 $0.1\times10^6m^3/h$，中非生产潜力为 $0.06\times10^6m^3/h$。目前，非洲规模化沼气工程数量很少，可开发的潜力巨大。

### 2.1.5 地热能资源

非洲地热能资源广泛分布在吉布提、埃塞俄比亚、肯尼亚、卢旺达、坦桑尼亚和乌干达等国家，但开发率非常低。潜在地热能主要源于东非大裂谷，目前已证实可开发量约 20GW。东非大裂谷（见图 2.5）地热带是全球 4 个高温地热带之一，其地热带主体位于非洲板块内，从红海往南，经过埃塞俄比亚高原后分东西两支，东支裂谷经过东非高原，西支裂谷经过坦噶尼喀湖，两支裂谷往南合并为马拉维裂谷往南延伸，以高热流、强烈现代火山作用以及广泛断裂活动为特征，热储温度多高于 200℃。

图 2.5　东非大裂谷

## 2.2 非洲可再生能源稳步发展

### 2.2.1 水电发展情况

2019—2023 年非洲水电（不含抽水蓄能）装机容量从 32991MW 增长至 37082MW，近五年增长了 12.4%（见图 2.6）。2023 年非洲水电装机容量排名前五的国家分别为埃塞俄比亚（4883MW）、安格拉（3729MW）、刚果民主共和国（3172MW）、赞比亚（3165MW）、尼日利亚（2851MW），如图 2.7 所示。

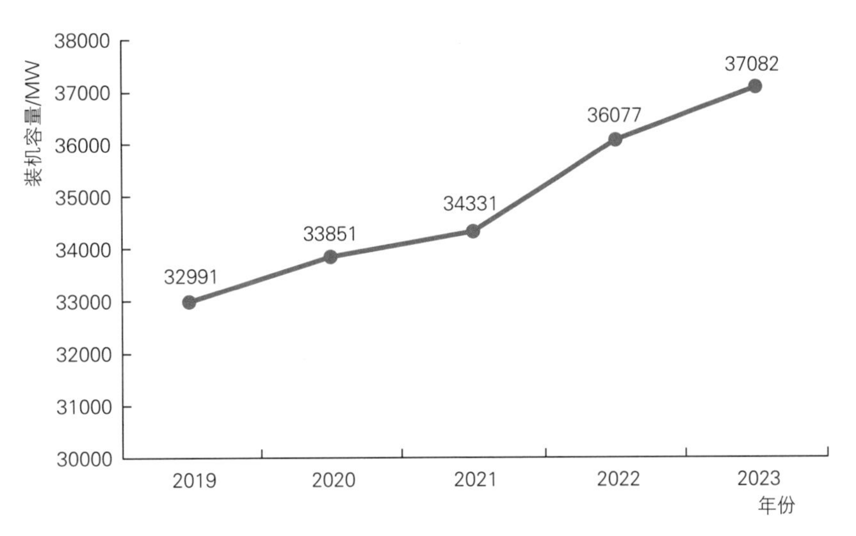


图 2.6 2019—2023 年非洲水电装机容量变化趋势（数据来源：IRENA）

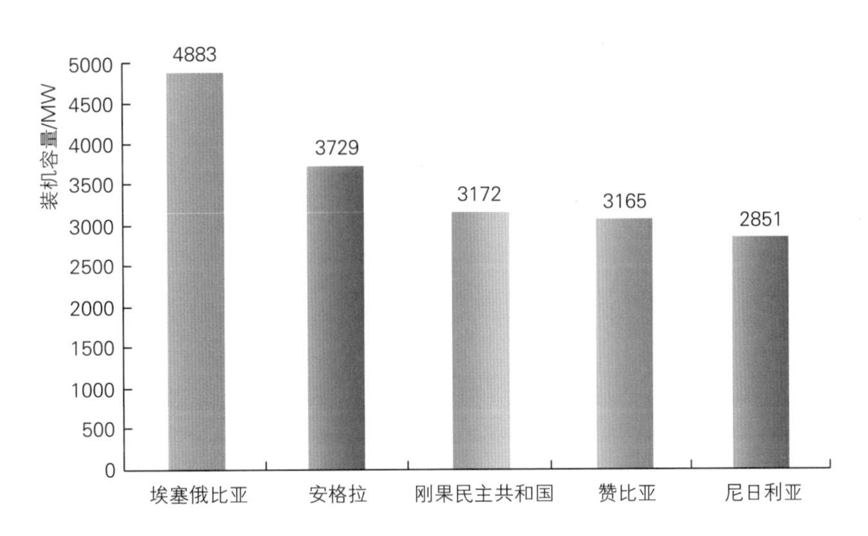


图 2.7 2023 年非洲水电装机容量排名前五国家（数据来源：IRENA）

### 2.2.2 风电发展情况

2019—2023 年非洲风电装机容量从 5528MW 增长至 8654MW，近五年增长了 56.5%（见图 2.8）。2023 年非洲风电装机容量排名前五的国家分别为南非（3442MW）、埃及（1890MW）、摩洛哥（1858MW）、肯尼亚（436MW）、埃塞俄比亚（324MW），如图 2.9 所示。

### 2.2.3 太阳能发展情况

2019—2023 年非洲太阳能发电装机容量从 9412MW 增长至 13479MW（其中光伏发电 12394MW、光热发电 1085MW），近五年增长了 43.2%（见图 2.10）。2023 年非洲太阳能发电装机容量排名前五的国家分别为南非（6164MW）、埃及（1856MW）、摩洛哥（934MW）、突尼斯（506MW）、阿尔及利亚（451MW），如图 2.11 所示。

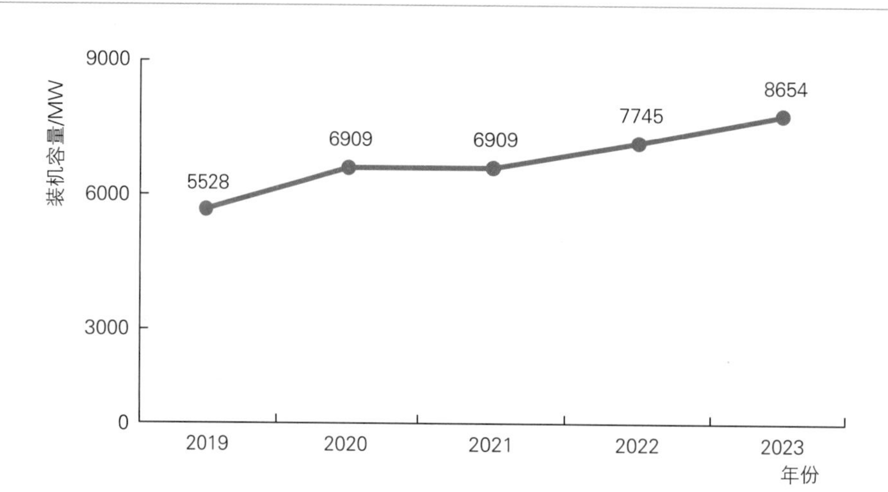


图 2.8　2019—2023 年非洲风电装机容量变化趋势（数据来源：IRENA）

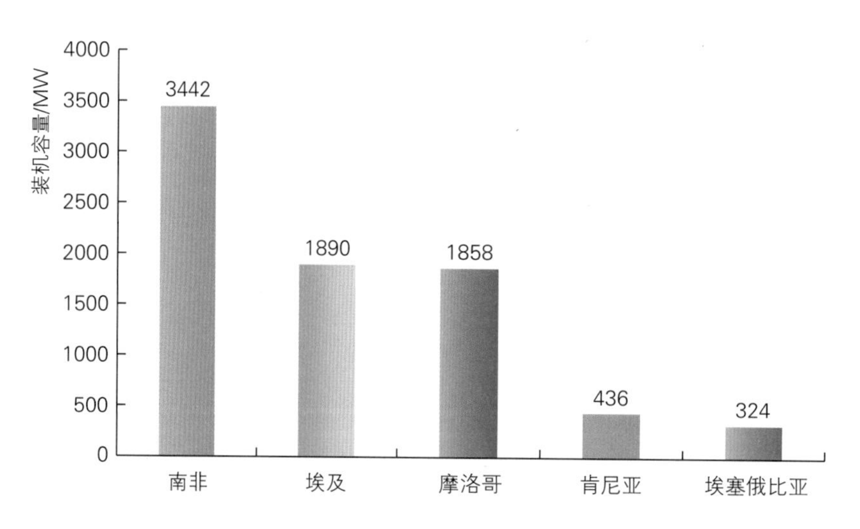


图 2.9　2023 年非洲风电装机容量排名前五国家（数据来源：IRENA）

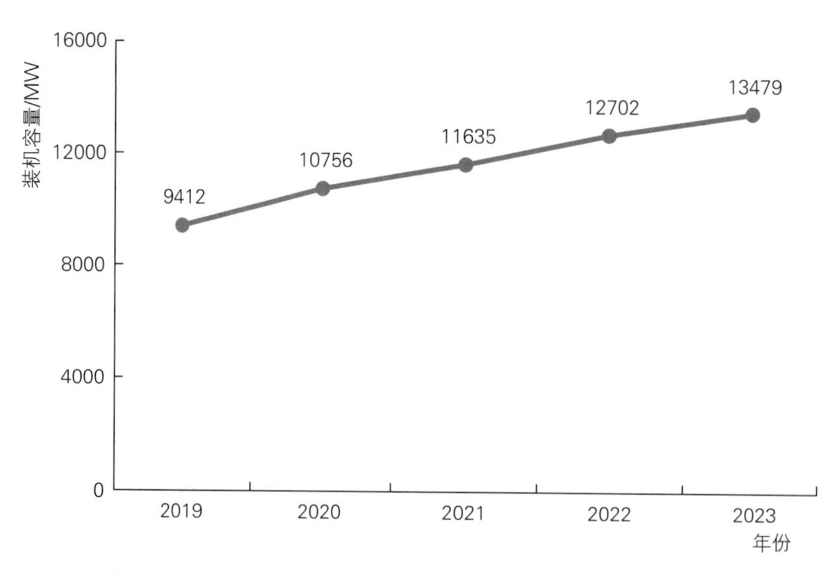


图 2.10　2019—2023 年非洲太阳能发电装机容量变化趋势（数据来源：IRENA）

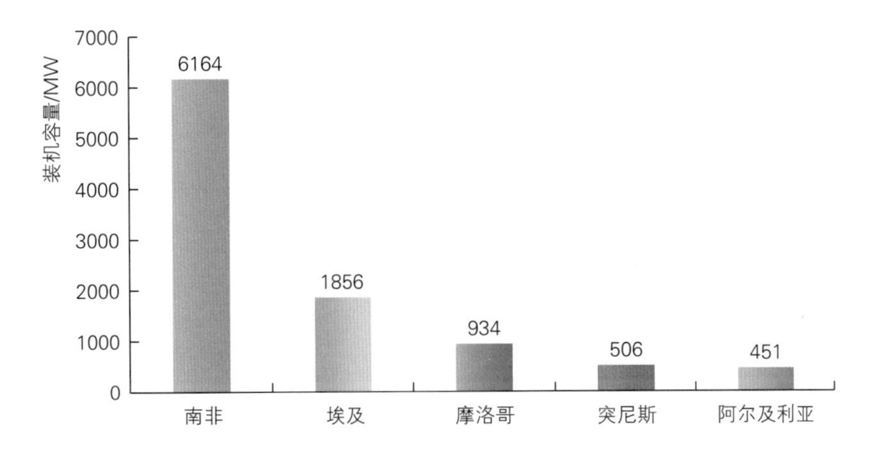


图 2.11 2023 年非洲太阳能发电装机容量排名前五国家（数据来源：IRENA）

### 2.2.4 生物质能发展情况

2019—2023 年非洲生物质能发电装机容量从 1807MW 增长至 1901MW，近五年增长了 5.2%（见图 2.12）。2023 年非洲生物质能发电装机容量排名前五的国家分别为埃塞俄比亚（310MW）、南非（265MW）、苏丹（199MW）、埃及（131MW）、斯威士兰（106MW），如图 2.13 所示。

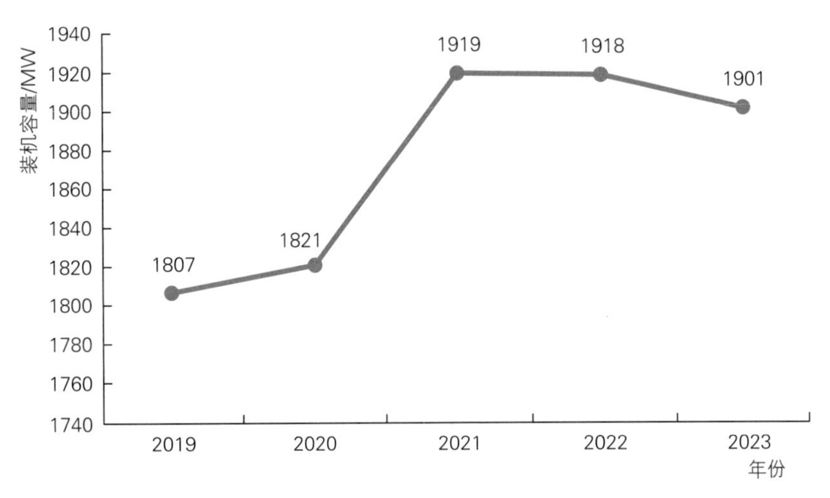


图 2.12 2019—2023 年非洲生物质能发电装机容量变化趋势（数据来源：IRENA）

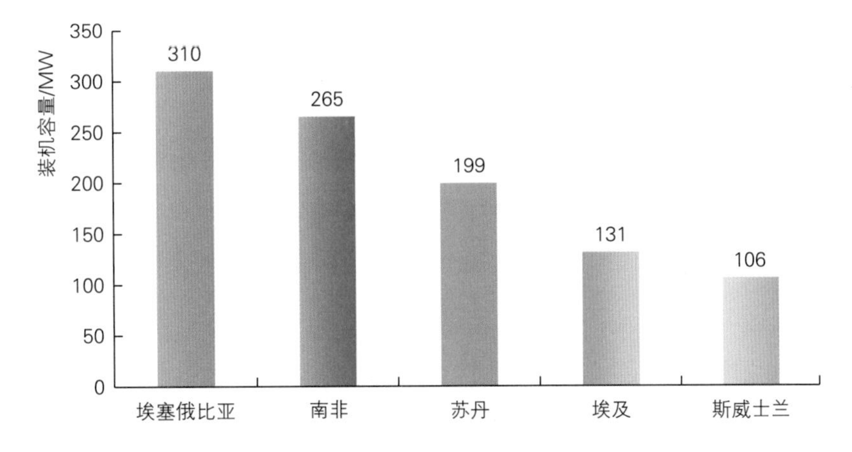


图 2.13 2023 年非洲生物质能发电装机容量排名前五国家（数据来源：IRENA）

### 2.2.5 地热能发展情况

2019—2023 年非洲地热能发电装机容量从 691MW 增长至 991MW，近五年增长了 43.4%（见图 2.14）。其中，肯尼亚地热能发电装机容量为 984MW，埃塞俄比亚地热能发电装机容量为 7MW。

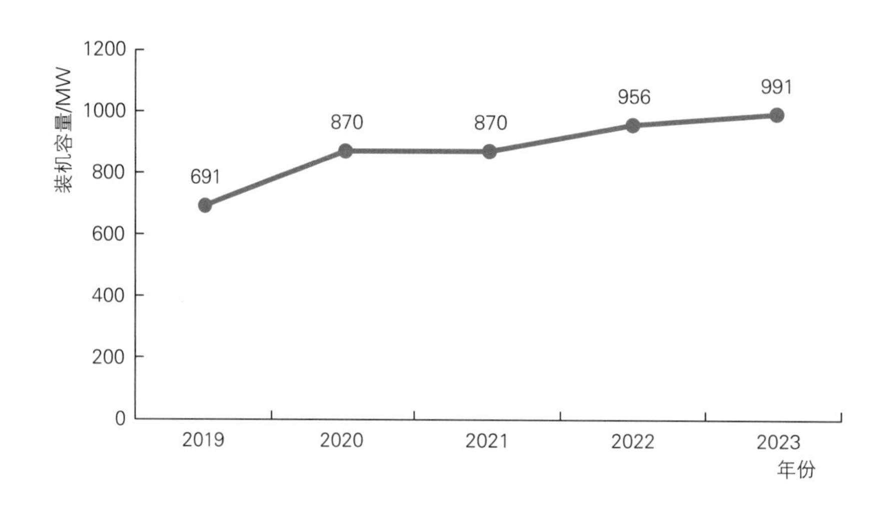


图 2.14　2019—2023 年非洲地热能发电装机容量变化趋势（数据来源：IRENA）

### 2.2.6 氢能发展情况

非洲拥有丰富的可再生能源资源，可再生能源制氢优势明显。据相关资料统计，截至 2023 年，非洲至少有 12 个国家正在开展氢能项目建设，19 个国家针对氢能产业制定了专门的监管框架或国家战略。非洲绿色氢能联盟研究显示，非洲北部和南部国家更加适合可再生能源制氢，预计到 2050 年，北非有 1200 万～3000 万 t 氢当量的出口机会，主要向欧洲输送氢气；南部非洲（以纳米比亚和南非为主）有 1000 万～2200 万 t 氢当量的出口机会。随着产能成本的下降和非洲电解槽供应链的扩大，2050 年非洲氢能生产成本有望降至不到 1.3 美元/kg。

❖ 德勤公司研究显示，北非、撒哈拉以南非洲、南美和中东等地区是全球绿氢潜力最大的地区，预计到 2050 年，这四个地区的氢能产量将占全球总量的 45%。

❖ 马斯达尔公司认为，到 2050 年，非洲丰富的太阳能和风能资源每年可生产 3000 万～6000 万 t 绿氢，为非洲创造 190 万～370 万个就业岗位。

❖ 欧洲投资银行研究指出，非洲有能力实现每年 1 万亿欧元的绿氢产值，并预测到 2035 年，非洲大陆每年可以生产 5000 万 t 绿氢，生产成本不到 2 欧元/kg。

### 2.2.7 储能发展情况

非洲风电和太阳能发电装机容量近五年（2019—2023 年）来分别增长了 56.5% 和 43.2%，由于风能、太阳能发电具有随机性、波动性和间歇性，未来储能的需求也不断加大。许多非洲国家电力系统不稳定，缺电限电严重，发展储能系统也成为保障家庭和工业电力供应的有效途径。根据全球能源互联网发展合作组织预测，2050 年可再生能源大规模开发利用将为非洲带来约 210GW、1230 GWh 的储能需求。

非洲正在积极推进储能发展。 2023 年非洲抽水蓄能装机容量为 3196MW，约占全球抽水蓄能总装机容量的 2.3%，近五年（2019—2023 年）抽水蓄能装机容量未有明显增长，但较十年前（2014 年）增长了 71.5%。 目前发展抽水蓄能的非洲国家有南非和摩洛哥，装机容量分别为 2732MW 和 464MW。 此外，据有关预测，2024—2029 年期间，非洲电池储能系统市场预计将以超过 5.2%的复合增长率增长。 南非电池储能发展较为突出，2023 年 11 月，南非宣布非洲大陆最大的电池储能系统项目落成，根据世界银行预测，2030 年南非电池存储市场将增长至 9700MWh。

### 2.2.8 小结

2023 年非洲发电装机总量为 252.8GW，化石能源仍是非洲当前的主力电源，约占非洲发电总装机容量的 3/4；可再生能源中，水电（不含抽水蓄能）装机容量为 37.1GW，约占全球水电装机容量的 3%；风电和太阳能发电装机容量分别为 8.7GW 和 13.5GW，在全球占比均不足 1%，未来非洲可再生能源发展潜力巨大。

非洲正在加快能源转型，推动可再生能源的开发。 如图 2.15 所示，自 2012 年起，非洲可再生能源（不含抽水蓄能）发电装机增速明显高于化石能源发电装机增速。 近五年（2019—2023 年），非洲可再生能源（不含抽水蓄能）总装机容量增长率达 23.2%，较近五年化石能源装机容量增长率（6.4%）高出了 16.8 个百分点。 总体上看，非洲可再生能源资源丰富、开发潜力大、能源转型意愿强烈，为发展可再生能源奠定了良好的基础。

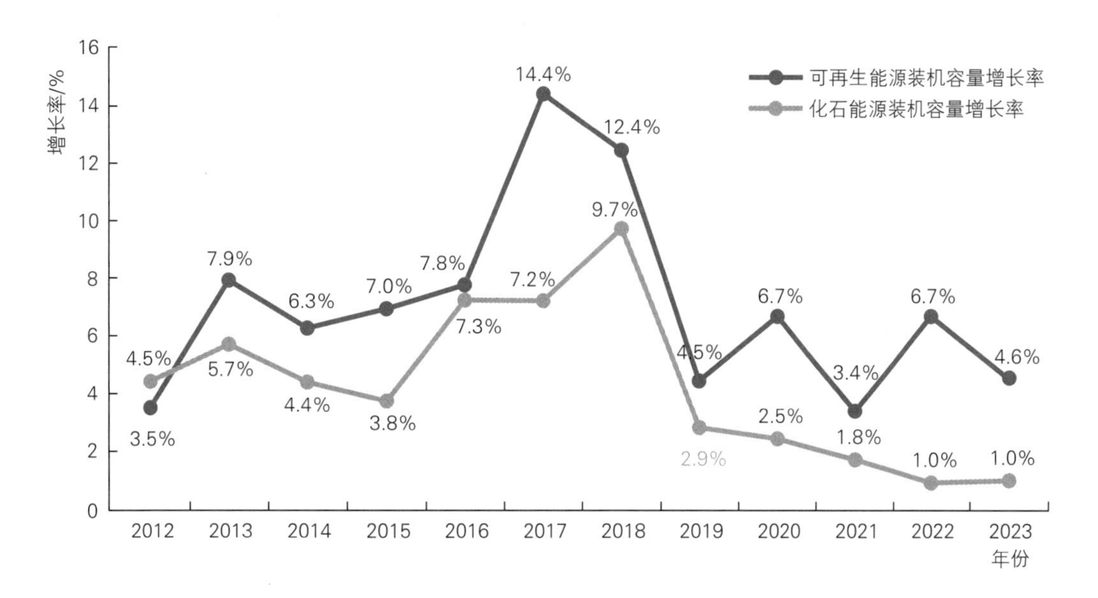


图 2.15 2012—2023 年非洲可再生能源（不含抽水蓄能）和化石能源装机容量增长率对比图（数据来源：IRENA）

## 2.3 非洲积极推动能源绿色发展

**《非洲基础设施发展计划》**

基础设施建设是非洲发展的基石，2012 年 1 月，由非盟委员会、“非洲发展新伙伴关系”以及非洲开发银行联合发起的《非洲基础设施发展计划》（PIDA）于非盟第十八届首脑会议上获得通过。PIDA 旨在充分调动相关资源，推动非洲跨境基础设施互联互通，全面实现基础设施现代化。

PIDA 预计于 2040 年完成，建设总投资 3600 亿美元，涵盖了能源在内的 4 个重点领域，其中能源领域的重点是发展高效、可靠、可负担得起且环境友好的能源系统，增加获得现代能源服务的机会，把加快可再生能源生产和传输作为重要内容，并利用可再生能源的快速发展为基础设施建设提供新的途径。

**《2063 年议程》**

2015 年 1 月，由非盟领导人发起的《2063 年议程》在第二十四届非盟峰会上获得通过。《2063 年议程》旨在为非洲未来 50 年发展规划制定行动纲领，体现了非洲国家和人民注重发展、期待繁荣、追求幸福的美好愿望，为充满生机和活力的非洲描绘出了一幅宏伟的蓝图，计划在 2013—2063 年的 50 年内实现其包容性和可持续发展目标，建成地区一体化、和平繁荣新非洲。

可再生能源是《2063 年议程》优先发展领域，到 2063 年，非洲将被全球公认为一个尊重环境、基于可持续发展和可再生能源、具有生态意识的大陆。非洲将充分发挥其能源生产潜力，加速非洲从传统能源向现代可再生能源的过渡，确保大多数公民的用电需求得到满足，促进经济增长和消除能源贫困，提高可再生能源占比。

**《非洲可再生能源倡议》**

2015 年 12 月，非盟在第 21 届联合国气候变化大会（COP21）框架下创立了《非洲可再生能源倡议》，旨在落实《巴黎气候变化协定》要求，提高非洲电气化率，加速开发现代化的可再生能源、提高能源的可及性和普惠性，让更多的非洲人民可以使用可再生能源，改善非洲人民福祉，推动非洲可再生能源革命，减少温室气体排放，使非洲国家走上可持续和气候友好型发展的道路。

《非洲可再生能源倡议》促进非洲国家跨越式发展可再生能源系统，支持低碳发展战略，同时加强经济和能源安全。通过确保人人都能获得足够的可再生、适当和负担得起的能源，助力实现经济可持续发展。到 2030 年，在非洲大陆新增 300GW 的可再生发电能力，以水电、太阳能发电为主，辅之以地热发电、风电和生物质发电。

**非洲大陆统一电力系统**

非盟在 2018 年提出了构建覆盖整个非洲大陆统一电力系统（AfSEM）的愿景，旨在通过电力基础设施的互联，整合非洲各国电力市场，促进电力贸易和互联互通的倡议，提高非洲的能源安全，促进可再生能源的开发和利用，以及支持非洲的经济发展。AfSEM 根于非盟《2063 年议程》，是非洲世界级综合基础设施重要组成部分。AfSEM 的建立将全面实现非洲电力可及，并进一步促进非洲可持续发展。

AfSEM 的目标是为非盟成员国提供更高层次的能源安全、可持续性与竞争力。 它将成为全球最大的单一电力市场，覆盖 55 个成员国，为近 14 亿人口提供服务。 AfSEM 将有效应对非洲日益增长的电力需求，以最具成本效益的方式进行。 它不仅是开发非洲大陆可再生能源潜力的关键工具，也是推动非洲实现全面电力覆盖的有力加速器，为非洲各区域合理利用电力资源提供便利的同时，也为非洲绿色发展提供助力。

**《非洲大陆电力系统总体规划》**

为加快 AfSEM 的推进，非盟在 2020 年委托非洲联盟发展署牵头启动《非洲大陆电力系统总体规划（CMP）》编制工作，经过三年的深入研究和细致规划，CMP 于 2023 年 9 月 15 日获得了正式批准。CMP 是 AfSEM 的有力支撑，为全面实现非洲电力可及和电力互联提供了路径方案。

CMP 深入分析了非洲电力系统的现状、未来需求、资源评估、计划实施的合适途径、整个非洲大陆的贡献、投资需求以及为实现所需能源输送和贸易所需的基础设施。 预计若要 2035 年全面实现非洲电力可及的目标，到 2040 年非洲可再生能源装机将达到 750GW，2023—2040 年需要的电力投资约为 1.3 万亿美元。

**《非洲领导人关于气候变化的内罗毕宣言及行动呼吁》**

《非洲领导人关于气候变化的内罗毕宣言及行动呼吁》（以下简称《内罗毕宣言》）于 2023 年 9 月 6 日在首届非洲气候峰会上发布。《内罗毕宣言》以非盟《气候变化与韧性发展战略和行动计划（2022—2032）》为基础，表达了非洲应对气候变化的集体愿望，构成了非洲参与第 28 届联合国气候变化大会（COP28）的基础。

《内罗毕宣言》指出，非洲拥有巨大可再生能源发展潜力以及丰富的自然资源，重申非洲愿意创造有利环境，制定政策并促进必要的投资，以释放可再生能源潜力，履行非洲的气候承诺，并为全球经济脱碳作出有意义的贡献。 非洲国家领导人在宣言中建议国际社会协助非洲提升可再生能源发电能力，从 2022 年的 56GW 提升至 2030 年的 300GW，该目标高于“阿联酋共识”中的到 2030 年增加至 2022 年 3 倍的要求。

**“沙漠发电”倡议**

该倡议由非洲开发银行领导，目标是利用萨赫勒地区丰富的太阳能资源，将其转变为可再生能源发电站。 该计划涵盖了布基纳法索、乍得、吉布提、厄立特里亚、埃塞俄比亚、马里、毛里塔尼亚、尼日尔、尼日利亚、塞内加尔和苏丹等 11 个国家，旨在到 2030 年通过公共、私人以及并网和离网项目，实现 10GW 的太阳能发电装机容量，从而为 2.5 亿人提供电力。

除了太阳能发电，该计划还包括区域电力贸易，如毛里塔尼亚-马里电力互联项目，旨在提升区域能源互通和电力普及率。 此外，倡议得到世界银行非洲可持续能源基金的支持，以促进私营部门参与。 这不仅有助于减少对化石燃料的依赖，降低温室气体排放，还是非洲向可持续能源转型的重要一步，对减少森林砍伐和保护生态环境具有重要意义。

**非洲可再生能源走廊**

非洲可再生能源走廊（ACEC）是一项区域性倡议，旨在加速开发可再生能源潜力，并在东部非洲电力池（EAPP）和南部非洲电力池（SAPP）内促进可再生能源的跨境贸易。该倡议建立在非洲领导人加强区域机构和输电基础设施的政治承诺之上，旨在形成大型有竞争力的市场，降低各生产部门的成本。

通过创建更大的区域电力市场，ACEC 可以在 2030 年前吸引投资，满足 EAPP 和 SAPP 地区 40%～50%的电力需求。ACEC 在 2030 年前需要每年投入高达 250 亿美元用于发电，另外还需要每年投入 150 亿美元用于电网基础设施。共同努力将使资源多样化，改善能源安全，并创造投资机会和就业增长。扩大可再生能源应用也为非洲提供了一个全面的机会，可以避免碳密集型基础设施锁定，并迅速向低碳未来迈进。

### 非洲小型电网项目

非洲小型电网项目（AMP）（2022—2027 年）是一个由各国政府主导的技术援助项目，总额为 5000 万美元，由全球环境基金（GEF）资助，由联合国开发计划署联合落基山研究所（RMI）、非洲开发银行（AfDB）合作实施，旨在刺激太阳能电池小型电网市场，以提高 21 个国家的电力接入。该项目的目标是通过专注于支持能源的生产性用途，将能源接入带来的发展益处带给非洲大陆的众多社区，从而促进农业、卫生保健、教育和小型企业等需要能源投入的部门的质量提升，从而促进社会经济的发展。

21 个国家包含安哥拉、贝宁、布基纳法索、布隆迪、乍得、科摩罗、刚果民主共和国、吉布提、埃塞俄比亚、斯瓦蒂尼、利比里亚、马达加斯加、马拉维、马里、毛里塔尼亚、尼日尔、尼日利亚、圣多美与普林西比、索马里、苏丹、赞比亚。

### 扩大太阳能应用计划

扩大太阳能应用计划由世界银行通过 IFC 发起，以招标的形式协助非洲国家安装光伏电站，其目标是鼓励私营公司投资光伏，向相关国家电网提供电力。扩大太阳能应用计划提供了一种标准化的项目开发方法，包括招标流程、融资和风险管理。赞比亚、塞内加尔和埃塞俄比亚等国家已成功实施了太阳能项目，大大降低了这些地区的太阳能发电成本。

# 3 中非可再生能源合作历程与成就

“中国和非洲国家已经有着多领域的合作，这为帮助非洲国家实现创新绿色发展打造了很好的基础。”

——联合国秘书长 安东尼 · 古特雷斯

自新中国第一代领导人与非洲老一辈政治家共同奠定友好基础以来，双方在相互尊重、热爱与支持中走出了一条合作共赢之路。 74 年来，中非双方风雨同舟、携手前行，中国致力于不断巩固中非政治互信，深化各领域务实合作，为非洲和平与发展提供力所能及的帮助，中国对非合作一直走在国际对非合作的前列。 在中非元首外交引领和顶层设计指引下，双方可再生能源领域合作机制日益完善，技术创新与时俱进，项目合作开花结果，合作模式从早期的对外援助、以工程承包为主，正在朝投建营一体化方向发展。 1964 年中国政府援建的几内亚金康水电站是中非合作的第一个可再生能源项目，过去几十年，中国企业积极响应“走出去”战略，在非洲建设了上百个可再生能源项目，包括科特迪瓦苏布雷水电站、南非德阿风电站、中非共和国萨卡伊光伏电站等重点标志性工程。 可再生能源领域合作成果遍布非洲大地，改善了非洲经济社会发展条件，给双方人民带来了实实在在的好处。

## 3.1 中非合作源远流长

新中国成立后，中非在反帝反殖、争取民族解放斗争中相互坚定支持，结下了牢不可破的兄弟情谊。 1955 年万隆会议上，周恩来总理提出的“求同存异”方针为中国与亚非国家间的团结合作奠定了基础。 随后中国与埃及于 1956 年 5 月建立外交关系，开启了中非官方交往的新纪元。 在此基础上，至 1963 年年底，中国与 12 个非洲国家建立了外交关系。 1963—1964 年，周恩来总理等中国领导人对非洲 10 国的访问，被誉为中非关系发展的“开山之旅”，进一步加深了双方的联系。 这些交往不仅标志着中非关系从民间到官方的转变，也为后续中非合作奠定了平等、真诚的基础，周恩来总理倡导的《和平共处五项原则》成为指导双方关系的核心原则，具有深远的国际意义。

在 20 世纪 70 年代，中国与非洲的友好合作关系不断加强，共与 25 个非洲国家建立了外交关系，并对非洲国家提供了大量的无偿经济援助，1965—1969 年间援助总额达到 2 亿美元，特别支持了坦桑尼亚、赞比亚和几内亚等国的发展。 1976 年，中国援建的坦赞铁路竣工，成为连接东非和中南非的重要交通干线，该项目是中国最大的援外项目之一，全长 1860. 5km。 1977 年，中国援建的索马里摩加迪沙体育场也顺利竣工，成为索马里的标志性建筑。 在这一时期，尽管中国国内面临诸多困难和挑战，但中非关系依然保持了良好的发展势头，中国继续支持非洲的民族独立和解放运动，而非洲国家则坚定支持中国在国际舞台上的地位，特别是支持中国恢复在联合国的合法席位。 这段历史见证了中非之间深厚的友谊和相互支持的伙伴关系。

改革开放后，中非关系迈入了以务实合作为特征的新阶段。20 世纪 80 年代，中国的劳务合作和承包工程在非洲的营业额超过 25 亿美元；90 年代，中非年均贸易额从 10 多亿美元增长至 1999 年的 64.8 亿美元。在这一时期，中国领导人访问非洲多国，宣布了与非洲国家开展经济技术合作的四项原则，即平等互利、讲求实效、形式多样、共同发展，这标志着中非经济技术合作进入了新阶段。同时，中非关系从无偿援建转为向贸易和基础设施建设合作，非洲成为中国实施多边外交的重要伙伴，双方经济互补性强，经贸合作迅速发展。

2000 年，中非合作论坛应运而生，引领中非友好务实合作实现跨越式发展，中非关系进入了机制化合作的新阶段。2000 年 10 月，首届论坛在北京召开，旨在构建国际政治经济新秩序，推动中非合作，通过了《中非合作论坛北京宣言》和《中非经济和社会发展合作纲领》。2006 年，胡锦涛访问非洲三国并在尼日利亚国会发表演讲，强调中非友好关系，并提出发展中非新型合作伙伴关系的五点建议。同年 11 月，北京峰会成功举办，是新中国成立以来规模最大的外事活动之一，通过了指导中非关系发展的重要文件。

2013 年，习近平就任国家主席后首次出访非洲，指出中非从来都是命运共同体，提出“真实亲诚”理念和正确义利观，为新时代对非合作指明了前进方向、提供了根本遵循。在 2015 年 12 月中非合作论坛约翰内斯堡峰会上，习近平主席提出将中非关系提升为全面战略合作伙伴关系。2018 年 9 月，习近平主席在中非合作论坛北京峰会上提出构建更加紧密的中非命运共同体。2021 年 11 月，习近平主席在中非合作论坛第八届部长级会议上提出携手构建新时代中非命运共同体。十多年来，习近平主席五次访非，两次出席中非合作论坛峰会，推动中非合作在“一带一路”倡议和中非合作论坛等框架下不断深化，主持召开中非团结抗疫特别峰会、中非领导人对话会，总结提炼中非友好合作精神，亲自擘画新时代中非关系发展蓝图，持续造福双方民众，书写了中非友好合作的新篇章。

中非合作论坛已走过 20 多年历程，成为引领中非合作乃至国际对非合作的一面旗帜。中非关系实现了从新型伙伴关系到新型战略伙伴关系再到全面战略合作伙伴关系的“三级跳”跨越式发展。从“十大合作计划”到“八大行动”再到“九项工程”，中非合作不断拓展升级，硕果累累，贸易额从 2000 年的 105 亿美元增至 2023 年的 2821 亿美元，增长近 26 倍。中国对非投资存量从 2000 年不足 5 亿美元，增长至目前超 400 亿美元。多年来，中非合作新建和升级近 10 万 km 公路、超 1 万 km 铁路、近千座桥梁、近百个港口。中非各领域交往空前活跃，旅游、文化、青年、媒体交流互鉴百花齐放，双方友好民意基础更加巩固。

## 3.2 中非可再生能源合作成果丰硕

### 3.2.1 机制建设不断完善

中非可再生能源领域机制建设日益完善。在“一带一路”倡议和中非合作论坛指引下，双方能源领域发展战略深度对接；在应对气候变化南南合作方面，中国积极参与应对气候变化全球气候治理；在联合国、二十国集团（G20）、亚太经济合作组织（APEC）等多边机制下，中国与国际社会携手巩固深化区域能源合作；在中国 - 非盟能源合作伙伴关系框架下，积极推进政策与信息交流、能力建设、三方合作等领域务实合作。截至 2021 年年底，中国已同 21 个非洲国家和非盟委员会建立双边委员会、外交磋商或战略对话机制，同 51 个非洲国家建立经贸联（混）合委员会机制，中国与非洲国家在上述双边合作机制下，推动可再生能源合作项目孵化落地。

### "一带一路"倡议下可再生能源合作结出共赢之果

2013 年 9 月和 10 月，习近平主席分别提出建设"新丝绸之路经济带"和"21 世纪海上丝绸之路"（"一带一路"）合作倡议。近十年来（2013—2023 年），中国与 150 多个国家、30 多个国际组织签署了共建"一带一路"合作文件，截至 2023 年年底，已有 52 个非洲国家以及非盟委员会同中国签署共建"一带一路"合作文件，非洲成为参与"一带一路"倡议最重要的大陆之一。在倡议指引下，双方建设了一批"一带一路"标志性绿色工程，有力促进了非洲有关国家能源产业发展，为其他发展中国家提供借鉴和示范。随着"一带一路"倡议深入推进，中非可再生能源领域政策沟通不断深化，合作共识进一步凝聚，非洲可再生能源资源加速整合，能源市场不断释放，发展规模日渐扩大。

2019 年 4 月 25 日，第二届"一带一路"国际合作高峰论坛期间，"一带一路"能源合作伙伴关系在北京成立，旨在推动能源互利合作，助力各国共同解决能源发展面临的问题，实现共同发展、共同繁荣，是画好"一带一路"工笔画的重要举措。目前，"一带一路"能源合作伙伴关系成员国数量达 33 个，其中包含 9 个非洲国家。2020 年，中国与非盟签署共建"一带一路"合作工作协调机制的谅解备

南非驻华大使在专访中表示：共建"一带一路"成果契合非洲发展需求

中非基础设施合作论坛项目签约仪式

忘录，非盟是首个同中国签署共建“一带一路”合作规划并率先建立工作协调机制的区域组织，双方在“一带一路”框架下开展的可再生能源等基础设施领域合作有效促进了非洲大陆区域经济一体化。

基础设施建设是共建“一带一路”倡议优先领域。2023 年 6 月，第三届中非基础设施合作论坛召开，论坛签署了对非工程承包、工程投资领域 19 个合约，涉及尼日利亚、肯尼亚、加纳、乌干达、科特迪瓦、刚果共和国、埃及、摩洛哥、津巴布韦、尼日尔、南非等多个非洲国家的能源、通信、工业及农业等专业领域的基础设施合作，签约项目总金额逾 29 亿美元，签约单位共计 32 家。

### 中非合作论坛框架下可再生能源合作不断深化

中非合作论坛成立于 2000 年 10 月，是中国和非洲国家之间在平等互利基础上的集体对话机制，旨在进一步加强中非在新形势下的友好合作，谋求共同发展。自中非合作论坛成立以来，双方聚焦风电、光伏发电等可再生能源产业发展和电力基础设施建设，有效推动非洲绿色转型与可持续发展，取得了一系列合作成果。在论坛机制下，中国已在非洲实施了上百个可再生能源和绿色发电项目，促进双方优势资源互补，推动实现互利共赢。

2015 年中非合作论坛约翰内斯堡峰会上提出中非“十大合作计划”，承诺“中方将支持非洲增强绿色、低碳、可持续发展能力，支持非洲实施 100 个清洁能源项目”。2018 年北京峰会通过了《关于构建更加紧密的中非命运共同体的北京宣言》和《中非合作论坛—北京行动计划（2019—2021）》两项成果文件，提出中非双方加强能源、资源领域政策对话和技术交流，对接能源、资源发展战略，开展联合研究，共同制定因地制宜、操作性强的能源发展规划。2021 年中非合作论坛第八届部长级会议通过了《中非合作论坛第八届部长级会议达喀尔宣言》和《中非合作论坛—达喀尔行动计划（2022—2024）》两项成果文件，提出中国将同非洲国家在中国-非盟能源伙伴关系框架下加强能源领域务实合作，共同提高非洲电气化水平，增加可再生能源比重，逐步解决能源可及性问题，推动双方实现能源可持续发展。中国将加强同非盟和非洲国家发展战略对接，支持非方加快落实非盟《2063 年议程》，早日实现自主可持续发展。

习近平主席在 2018 年中非合作论坛北京峰会上宣布设立中国－非洲经贸博览会。2023 年 6 月，第三届中国－非洲经贸博览会召开，以“共谋发展，共享未来”为主题，贯彻落实中非合作论坛

2018 年中非合作论坛北京峰会

2023 年第三届中国-非洲经贸博览会

第八届部长级会议精神。会上签约项目 120 个、金额 103 亿美元，发布 99 个对接合作项目、金额 87 亿美元，其中 11 个非洲国家发布 74 个对接项目，数量为历届之最。能源领域合作项目是重点签约对象之一，包括津巴布韦 100MW 光伏电站项目、尼日尔 250MW 风光储智慧能源园区项目、埃及塞得港绿氨项目、南非 500MW 光伏和 1000MWh 储能项目等。

### 多双边机制推动可再生能源合作走深走实

在联合国框架及 G20、APEC、金砖国家等重要多边机制下，中非共同推动绿色可持续发展。中非双方不断丰富和完善政府间对话、磋商及合作机制，充分发挥统筹协调作用，促进中非各领域合作全方位发展。在中国－摩洛哥能源合作执委会、中国－南非经贸协会能源委员会等能源领域双边合作机制下，中国与非洲重点区域及国别可再生能源合作走深走实。

中国长期以来高度重视提升非洲电力可及水平。早在 2015 年，中国作为 G20 创始成员国之一，坚定支持《G20 能源可及性行动计划：能源可及性自愿合作》，该计划将提升撒哈拉以南非洲地区电力可及性作为重要内容。2022 年 11 月，在 G20 领导人第十七次峰会上，中国率先公开支持非盟加入二十国集团倡议，推动国际社会进一步关注非洲电力可及问题，促进中非在 G20 等多边框架下开展可再生能源合作。在金砖国家领导人第十五次会晤期间，中国与南非签署《中南关于同意深化“一带一路”合作的意向书》，双方能源企业签署相关战略合作备忘录，积极参与南非可再生能源项目投资。2021 年 6 月 7 日，中国国家能源局与 IRENA 签订谅解备忘录，双方同意围绕能源转型战略与政策、可再生能源技术推广与应用，以及帮助和支持其他国家可再生能源发展三个方面开展合作；同年，IRENA 与相关组织合作支持非盟发展署制定 CMP，审查和重新考虑发电方案，以最大限度地提高社会经济效益，同时最大限度地减少排放。

2023 年 4 月，中国国家能源局应邀派团访问纳米比亚和安哥拉，就深化能源领域合作交换意见。在阿尔及利亚总统访华期间，中阿能源主管部门签署《中国-阿尔及利亚可再生能源合作谅解备忘录》，推动开展能力建设、联合研究、示范项目等合作。中非双方以各种形式在多双边机制下开展能力建设培训，为非洲各国培养了大量人才。据悉，仅 2015—2018 年，中国就为非洲培训了 20 万名各类职业技术人员，并提供了 4 万个来华培训的名额和 2000 个教育名额。据不完全统计，2018 年以

2023 年中国和南非两国能源主管部门领导会见

2024 年赞比亚国家执政能力建设与宏观经济规划部级研讨班

来，中国在吉布提、埃及、南非、肯尼亚、尼日利亚、科特迪瓦等 11 个非洲国家设立 12 所鲁班工坊，同非洲分享包括新能源在内的各领域优质教育，助力非盟实现“要让 70%的青年拥有一技之长”的目标。

### 积极推动应对气候变化南南合作

截至 2023 年年底，中国已与 41 个发展中国家签署 50 份气候变化南南合作谅解备忘录，其中与尼日利亚、埃塞俄比亚、贝宁等 16 个非洲国家签署了 18 份应对气候变化南南合作文件，通过合作建设低碳示范区、开展减缓和适应气候变化项目、组织能力建设培训等方式，为包括非洲国家在内的发展中国家提供支持。

2021 年中非合作论坛第八届部长级会议前，中非双方共同制定《中非合作 2035 年愿景》，作为愿景首个三年规划，中国将同非洲国家密切配合，共同实施包括“绿色发展”在内的“九项工程”。《中非合作 2035 年愿景》提出，中非能源合作向可再生、低碳转型，打造绿色发展新模式，实现中非生态共建。会上双方通过《中非应对气候变化合作宣言》，提出加强中非应对气候变化合作，在可再生能源等领域实施务实合作项目，共同应对气候变化挑战。《中非应对气候变化合作宣言》指出，愿进一步加强中非应对气候变化南南合作，拓宽合作领域，在可再生能源等领域开展务实合作项目。

为进一步落实《中非应对气候变化合作宣言》，中国在非洲气候峰会上宣布将开发实施应对气候变化的南南合作“非洲光带”项目，聚焦非洲光伏资源和可再生能源发展需求，利用中国光伏产业发展优势，采取“物资援助 + 交流对话 + 联合研究 + 能力建设”的方式，打造中非光伏资源利用合作示范带。 此外，中国还通过实施减缓和适应气候变化项目、共同建设低碳示范区等方式为非洲应对气候变化提供支持。 在 COP28 会议期间，中国与乍得签署首份“非洲光带”项目文件，与圣多美和普林西比就合作开展气候变化南南合作“非洲光带”项目签署谅解备忘录。

2021 年中国与塞舌尔签署应对气候变化南南合作低碳示范区建设实施的谅解备忘录

2023 年中国在非洲气候峰会上宣布将开发实施“非洲光带”项目

### 中国-非盟可再生能源合作取得积极进展

2018 年 4—5 月，中国国家能源局委托两个专家组分赴埃塞俄比亚、肯尼亚、南非、埃及、尼日利亚和科特迪瓦六个国家进行调研，为推动中非能源多层次、多角度的合作奠定了基础，助推中国与非盟及非洲国家建立政府间能源合作机制。 2018 年 9 月，中国国家能源局与非盟委员会签署《关于加强能源领域合作的谅解备忘录》，双方同意在“共商、共建、共享”的原则指导下，共同推动中国与非洲国家和非洲地区组织在可再生能源等领域开展全面合作。

2021 年 10 月，中国国家能源局与非盟委员会签署并成立中国-非盟能源合作伙伴关系（本段以下简称“伙伴关系”），双方同意在政策与信息交流、能力建设、项目合作、三方合作等领域开展务实合作。中非合作论坛第八届部长级会议通过的两项成果文件《中非合作论坛第八届部长级会议达喀尔宣言》和《中非合作论坛—达喀尔行动计划（2022—2024）》，均将伙伴关系作为中非能源领域务实合作的重要内容。中非双方依托伙伴关系已召开系列研讨会议、能力建设培训、项目推介等活动。

2018 年中国国家能源局委托专家组与埃及电力与可再生能源部交流

2022 年中国-非盟能源伙伴关系框架下首届能力建设培训期间考察大渡河安谷水电站

在促进交流研讨方面，通过系列研讨会议，中国和非洲国家政府部门、金融机构、能源企业交流技术经验，分享项目机遇，同时呼吁国际社会关注非洲可再生能源发展。在推动能力建设方面，联合有关单位，组织可再生能源和电网相关主题的能力建设培训，提升非洲能源行业从业者管理和技术水平。在深化项目合作方面，以“分享非洲能源发展机遇，推动中非项目务实合作”为主题召开项目合作推介会，旨在助力中国企业寻找非洲能源项目合作机遇，积极参与非洲基础设施发展规划（PIDA）项目及非洲国家能源项目投资建设。“小而美”示范项目建设方面，借助能源合作研讨会议、加速器项目和

2023 年 COP28 中国角宣布“中非能源创新合作加速器项目”正式上线

2023 年中国-非盟能源伙伴关系框架下首届能源合作项目推介会

COP28 相关活动，推广创新合作项目案例及技术解决方案，推动“小而美”项目落地非洲。

在第 27 届联合国气候变化大会（COP27）期间，中非有关单位依托中国–非盟能源伙伴关系发起《清洁能源提升非洲电力可及性埃及倡议》，呼吁国际社会进一步关注非洲地区无电人口用电问题，在促进多样性、公平性和包容性原则（DEI 原则）的基础上，采取更加有力和可持续的行动推动非洲可再生能源发展，为提升非洲电力可及贡献力量。

### 3.2.2 战略谋划持续推进

**《关于共同推进“一带一路”建设的合作规划》**

2020 年 12 月，国家发展和改革委员会主任与非洲联盟委员会主席签署《中华人民共和国政府与非洲联盟关于共同推进“一带一路”建设的合作规划》（以下简称《合作规划》），这是中国和区域性国际组织签署的第一个共建“一带一路”规划类合作文件，围绕政策沟通、设施联通、贸易畅通、资金融通、民心相通等领域，明确了合作内容和重点合作项目，提出了时间表、路线图。

《合作规划》的签署，将有效推动共建“一带一路”倡议同非盟《2063 年议程》对接，促进双方优势互补，共同应对全球性挑战，推进共建“一带一路”高质量发展，为全球合作创造新机遇，为共同发展增添新动力。随着规划的签署，中方将与非盟委员会建立共建“一带一路”合作工作协调机制，进一步推动《合作规划》实施落地。

**《中非合作 2035 年愿景》**

为进一步凝聚中非战略共识，深化中非全面战略合作伙伴关系，加强全面务实合作，做好中非合作的中长期规划，2021 年，中非合作论坛第八届部长级会议期间通过了《中非合作 2035 年愿景》。该愿景结合中国 2035 年远景目标、联合国 2030 年可持续发展议程、非盟《2063 年议程》及非洲各国发展战略，确定了未来十五年中非合作的总体框架，描绘了 2035 年中非各领域的合作前景。

《中非合作 2035 年愿景》表明双方共同构建转型增长新格局，实现中非产业共促，加强能源合作向可再生、低碳转型。中国支持非洲提高水能、核能等可再生能源利用比例，基于各国发展水平和能源需求，积极开发太阳能、风能、地热、沼气、潮流、波浪等可再生能源，通过分布式供电技术为非洲偏远地区提供稳定、可负担电力供应，支持光伏产业发展。

**“九项工程”——“绿色发展工程”**

习近平主席在中非合作论坛第八届部长级会议上宣布，作为《中非合作 2035 年愿景》愿景首个三年规划，中国将同非洲国家密切配合，共同实施包括绿色发展工程在内的“九项工程”。绿色发展工程提出，“中国将为非洲援助实施 10 个绿色环保和应对气候变化项目，支持‘非洲绿色长城’建设，在非洲建设低碳示范区和适应气候变化示范区。”中国重点在提供气变援助、加强绿色合作等方面与非洲开展合作，将为非洲国家援助实施绿色环保和应对气候变化项目，与非洲有关国家合作建设低碳示范区、适应气候变化示范区，实施减缓和适应气候变化南南合作项目。中国还将积极实施中非绿色创新计划，促进环境技术合作，开展应对气候变化能力建设合作，鼓励企业建设可再生能源项目。

**《中非应对气候变化合作宣言》**

气候变化是当前突出的全球性挑战，为共同应对气候变化，中非双方在中非合作论坛第八届部长级会议期间通过了《中非应对气候变化合作宣言》。《中非应对气候变化合作宣言》提出建立新时代中非应对气候变化战略合作伙伴关系，开启中非推进绿色低碳发展的新篇章，凸显了未来中非双方在南南合作和“一带一路”框架下，进一步深化应对气候变化交流合作的决心。

《中非应对气候变化合作宣言》表明双方一致倡导创新、协调、绿色、开放、共享的可持续发展，愿进一步加强中非应对气候变化南南合作，拓宽合作领域，在可再生能源等领域开展务实合作项目。中方是非洲可持续发展的坚定支持者，支持非洲国家更好利用可再生能源。中方将进一步扩大在光伏、风能等可再生能源产业项目的对非投资规模，助力非洲国家优化能源结构，实现绿色、低碳、高质量发展。

### 3.2.3 项目合作拓展深化

中非可再生能源合作日益紧密，双方成功合作了大量优质可再生能源电力项目。据 IEA 统计，2010—2015 年，中国企业作为非洲大陆最为主要的承包商之一，承建的各类项目发电装机容量约占撒哈拉以南非洲新增容量的 1/3；2010—2020 年期间，中国企业在撒哈拉以南的非洲已建或在建电力项目超 200 个，涉及非洲 37 个国家和地区，装机总量达到约 17GW，相当于撒哈拉以南非洲现有装机容量的 10%，其中可再生能源项目装机容量占比过半。

为进一步了解中非可再生能源项目合作情况和发展趋势，2023 年，受国家能源局委托，水电总院统计了 18 家中国主要能源企业近十年（2013—2023 年）在非洲大陆开展的可再生能源合作项目，共收集到项目 117 个，涵盖水电、电网、风电、太阳能、地热、生物质以及其他类型项目。

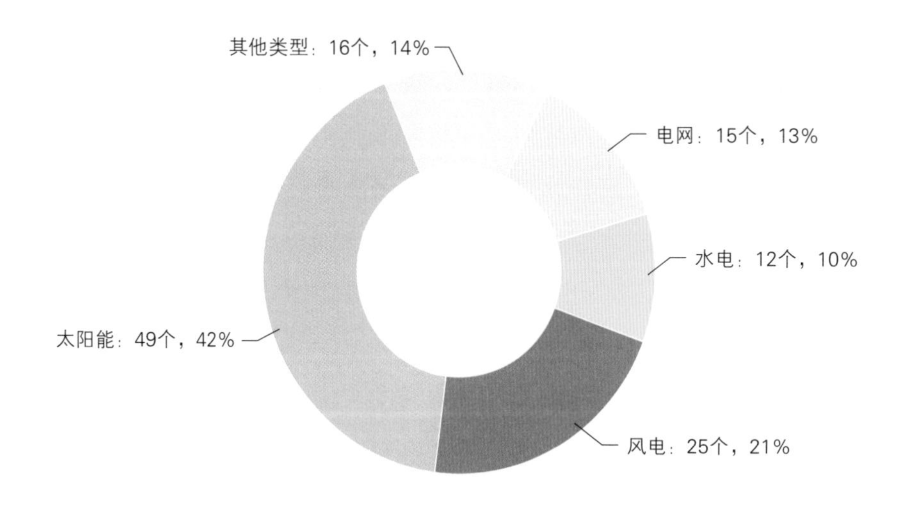


中非合作可再生能源项目统计

统计的 117 个项目中，超过 1/3 的项目已签约（含已建、在建和中标项目），总装机容量超过 10GW，主要集中分布在 7 个国家，分别是南非、加纳、埃及、科特迪瓦、尼日利亚、几内亚、赞比亚。上述 7 国项目数占统计项目总数近六成。目前还有约 19GW 项目正在积极开展前期工作。

科特迪瓦苏布雷水电站项目

## 典型水电项目案例

### 西部非洲——科特迪瓦苏布雷水电站项目

❖ 苏布雷水电站位于科特迪瓦西部，项目总装机容量 275MW，水库总库容 8300 万 $m^3$，总投资 5.6 亿美元。 项目为土石坝大坝，最大坝高 20m，坝线全长 4.5km。 中国企业负责项目 EPC 工作，承担枢纽工程、输变电工程等相关工作。 2017 年 5 月，电站首台机组并网发电，同年 11 月项目竣工，总工期 56 个月，比合同工期提前了 8 个月。

❖ 苏布雷水电站给科特迪瓦全国提供了约 14%的电力能源，有效保障了国家用电行业的稳定发展，改善了科特迪瓦电力能源结构，促进了可再生能源产业发展。 在工程建设过程中，项目雇用当地员工总数累计超过 5000 余人，属地化率达到 85.3%，并培养了一大批高级管理人员、施工技术人员、机组安装和电站运维人员，有力促进了当地就业。 项目先后分别获得中国驻科特迪瓦使馆颁发的“中非合作发展奖”，成为科特迪瓦国家能源平衡战略的核心项目。 科特迪瓦政府评价项目为“中科两国经贸合作的典范”，还授予项目中方负责人科特迪瓦“国家荣誉勋章”。

科特迪瓦总统出席项目竣工仪式

赞比亚下凯富峡水电站项目

## 典型水电项目案例

### 东部非洲——赞比亚下凯富峡水电站项目

✣　下凯富峡水电站位于赞比亚境内赞比西河支流卡富埃流域，总装机容量 750MW，于 2015 年 11 月开工建设；2021 年 6 月，首台机组并网发电；2023 年 3 月，5 台机组全部实现并网发电。

✣　下凯富峡水电站是赞比亚 40 年来开发规模最大的工程，被称为中赞合作的“一号工程”，也是“一带一路”典范工程。下凯富峡水电站对赞比亚供电稳定和可持续发展具有重要意义，截至 2023 年 11 月底，水电站累计发电量超 79.28 亿 kWh，提升赞比亚全国电力供应约 38%，减少二氧化碳排放约 706.2 万 t。中国企业通过建设学校、打井、修路架桥，支持当地教育、卫生、基础设施等发展。项目的实施为当地创造了约 1.5 万个就业岗位，培养了 160 余名技术人才。项目永久工程包括业主的运营村、永久发电工程等，寄宿中学是运营村工程中最核心的功能建筑。

项目配套建设的寄宿中学

安哥拉凯凯水电站项目

## 典型水电项目案例

### 南部非洲——安哥拉凯凯水电站项目

✣ 凯凯水电站位于安哥拉“母亲河”宽扎河上，距离首都罗安达约 230km，总装机容量 2172MW，是目前中资企业在安哥拉承建的最大水电站，也是非洲在建的最大水电工程，被誉为“非洲三峡工程”。 2023 年 5 月，在安哥拉总统的见证下，凯凯水电站成功完成大坝截流，掀开了大坝主体工程施工序幕。 2024 年，凯凯水电站临建工程基本上已经全部完成，正在进行大坝、厂房开挖等工作。

✣ 安哥拉的电力基础设施并不完善，包括首都罗安达在内的许多城市长期面临缺电问题。 凯凯水电站建成后，年平均发电量将达到 85.66 亿 kWh，将满足安哥拉全国 50%以上供电需求，每年减少温室气体排放量约 720 万 t，减少不可再生的石油和煤炭资源消耗 273.3 万 t，并大幅改善水资源利用条件，兼具调峰、防洪功能，助力当地经济社会发展。 自凯凯水电站开工建设以来，项目部还对当地员工开展培训工作，助力提升当地员工技能水平。 在水电站建设和维护过程中，项目部为当地民众提供了近万个工作岗位。

项目施工现场

南非德阿风电站项目

## 典型风电项目案例

南部非洲——南非德阿风电站项目

❖ 德阿风电站项目位于南非北开普省德阿镇附近，是中国发电企业在非洲首个集投资、建设、运营为一体的风电项目，安装中国自主生产的 1.5MW 风电机组 163 台，总装机容量 244.5MW，总投资额 25 亿元人民币。 项目于 2015 年 10 月开工；2016 年 9 月，首台风电机组顺利完成吊装；2017 年 8 月，163 台风电机组吊装、调试任务全部完成；同年 11 月，项目投产发电。

❖ 项目每年向南非电网输送约 7.6 亿 kWh 可再生电力，还为当地提供了 700 多个就业岗位，为改善当地民生和促进经济社会发展提供了新机遇。 中国企业积极融入当地，在德阿镇成立了 4 所儿童早教中心，为贫困家庭孩子提供教育平台，每年投入 450 万兰特资助大学贫困生完成学业。 截至 2023 年年底，已有 112 名成绩优秀的贫困大学生获得了奖学金项目资助。 同时，还设立了专项社区基金，并捐赠了医疗巴士，每年为超过 9000 多名社区成员提供免费的医疗服务。

中国企业成立的儿童早教中心

中非萨卡伊光伏电站项目

## 典型太阳能项目案例

### 中部非洲——中非萨卡伊光伏电站项目

❖ 萨卡伊光伏电站位于中非首都班吉，是中非的第一座光伏电站。2018 年 9 月，中非合作论坛北京峰会期间，中国和中非就中国援建光伏电站项目达成共识，被中非政府列为双边合作龙头项目。项目由中国企业设计施工，于 2021 年 4 月正式开工，项目总装机容量 15MW，配置 5MWh 储能系统，设计年总发电量 7600 万 kWh，于 2022 年 6 月成功并网发电。

❖ 班吉主要依靠柴油和水力发电，柴油成本高，水电发展慢，而光伏电站项目建设周期短，且绿色环保，有效解决了当地用电短缺问题。项目发电当年，电站满足了班吉市约 30% 的用电需求。此外，项目在建设过程中还提供了约 700 人次的就业机会，为当地社区培训了一批光伏电站建设人才，助力当地工人掌握了各项技能。"如今社区里有人新开了小卖部，有人新开了小饭馆，夜里照常营业。光伏电站让社区变得有生气，我相信以后还会更好。"住在班吉郊区的光伏电站线路安装工人吉拉贝赞叹道。

当地员工参与项目建设

摩洛哥努奥光热电站三期项目

## 典型太阳能项目案例

### 北部非洲——摩洛哥努奥光热电站三期项目

✣ 摩洛哥努奥 150MW 光热电站三期项目位于摩洛哥东南部瓦尔扎扎特市努奥太阳能发电园区，属于大容量塔式熔盐光热电站项目，镜场面积约 132 万 $m^2$，吸热塔高度 243m，储热时长 7.5h，共使用 7400 面巨型定日镜，年设计输送可再生电力 5.3 亿 kWh。项目在世界上首次采用混凝土和钢结构混合式结构的光塔，还应用了由中国企业研发的系列新技术、新工艺、新材料。超过 100 万户当地居民受惠。

✣ 项目采用 EPC 总承包方式建设，由中国企业负责设计、施工和设备供货。项目于 2015 年 5 月开工建设，2018 年 10 月竣工。项目先后获评摩洛哥“五星质量奖”“五星安全奖”“社会贡献奖”和“经济就业促进奖”等奖项，电站建设过程中，大量引进当地分包商、供货商等达 60 余家，周边辐射带动上百家当地企业参与，累计为当地创造了近 1.4 万个就业岗位。

电站冷热熔盐罐

埃塞俄比亚莱比垃圾发电项目

## 典型生物质项目案例

### 东部非洲——埃塞俄比亚莱比垃圾发电项目

❖ 莱比垃圾发电厂位于埃塞俄比亚首都亚的斯亚贝巴，总装机容量 50MW，包括 2×600t/d 城市垃圾焚烧炉排炉及辅助系统、2×25MW 凝汽式汽轮发电机组及辅助系统、1 座 132kV 单母线升压站。 项目采用国际领先的高性能垃圾焚烧技术和设备，烟气排放达到欧盟 2000 标准，设计处理垃圾量 1280t/d，年处理垃圾量 43.75 万 t，年发电量 1.85 亿 kWh。 项目于 2014 年 9 月正式开工，2017 年 4 月进入全面调试阶段，2017 年 9 月建成商运。

❖ 莱比垃圾发电厂是中国和埃塞俄比亚在环保领域合作的第一个项目，也是非洲首座垃圾发电厂。 项目建成后，一方面可以解决环境污染，另一方面，提高了当地电力可及水平。 莱比垃圾发电厂已成为中埃合作的亮点，也是中非绿色合作的标志性工程。 埃塞俄比亚电力公司表示："这是埃塞乃至非洲的标志性工程""该厂处理垃圾的意义要大于发电，因为像亚的斯亚贝巴这样的大城市，处理生活垃圾成为越来越关键的民生问题，有着非同寻常的社会意义。"

垃圾发电厂施工现场

埃塞俄比亚索马里州离网光伏电站项目

## 典型微网项目案例

### 东部非洲——埃塞俄比亚索马里州离网光伏电站项目

> “中国企业承建的 4 个光伏电站，是埃塞第一批 12 个离网太阳能电站中最早完工和投运的工程，最早为所在村落送去光明，我们要为中国建设者的坚守点赞，这是中国企业实力与担当的体现。”
>
> ——埃塞俄比亚水利能源部部长（时任）塞拉西 · 贝克利

✣ 索马里州离网光伏电站项目位于埃塞俄比亚索马里州科利尔村附近，电站设计年发电量为 142 万 kWh，总投资约 1400 万美元。 项目采用智能离网光伏系统，可以远程监测系统运行数据并调整运行参数。 基于分层分布式设计的微网控制系统可实现毫秒级的快速协调控制，配以业内领先的储能变流器和光伏逆变器，保障了微网系统的持续安全运行。 项目采用的微网能量管理系统、协调控制器、储能变流器和光伏逆变器等核心设备均为中国自主产品。

✣ 项目惠及了周边 2000 多户家庭，使近 6000 人用上了可再生电力能源。 当地村民表示：“今天是值得纪念的一天，漆黑的夜晚从此将被电灯照亮，孩子们可以在电灯下读书，我们终于有了梦寐以求的电。”世界银行表示，中国企业建造的该项目为“点亮非洲”计划作出很好示范，世界银行将以此为模板，持续在埃塞俄比亚其他 200 余个未通电村庄大力推广，力求在 2025 年实现全国 35% 的供电由离网光伏发电完成，惠及 570 万户家庭。

# 4 中非可再生能源合作展望

> 非洲是充满希望的大陆，我们对非洲的前景充满信心，21 世纪必将见证中国和非洲的共同发展振兴。
>
> ——中华人民共和国中央政治局委员、外交部部长王毅

> 我们珍视中方为非洲一体化、互联互通、自贸区建设提供的强有力支持，期待同中方一道，携手推进新时代中非命运共同体建设。
>
> ——非盟委员会主席穆萨 · 法基 · 穆罕默德

随着全球对可再生能源需求的增加，加快发展可再生能源成为全球共识。非洲经济增长韧性十足，用能需求不断释放，中国产业技术持续发展，互补优势不断扩大，中国与非洲的经贸合作高质量发展，为双方都带来了新机遇，中非可再生能源合作正当其时。但与此同时，中非双方也面临合作不确定性、资金短缺、基础设施与人才资源亟须加强等问题。在此背景下，中国应加强同非洲国家相互支持和友好合作，把握可再生能源合作的新机遇，共同应对风险挑战。展望未来，中非将持续深化可再生能源合作，通过战略对接、机制建设、政策交流和项目合作等方式，巩固和深化中非战略伙伴关系，推动双方能源转型和可持续发展，在现代化道路上携手向前，共筑高水平中非命运共同体。

## 4.1 把握机遇共迎挑战

### 4.1.1 合作机遇

**应对气候变化　发展可再生能源成为全球共识**

积极发展可再生能源，推动经济社会绿色低碳转型，已经成为国际社会应对气候变化的普遍共识。2022 年 12 月举行的 COP28 上，118 个国家承诺，到 2030 年，将全球可再生能源产能提高到 2022 年的 3 倍，将能源效率提高到 2023 年的 2 倍。联合国研究数据显示，2022—2030 年，非洲因气候变化遭受的损失预计最高可达 4405 亿美元。2023 年 9 月，首届非洲气候峰会通过《内罗毕宣言》，非洲各国元首在宣言中呼吁发展中国家和发达国家携手降低温室气体排放，并敦促发达国家兑现相关的出资和技术援助承诺，建议国际社会协助非洲提升可再生能源发电能力，从 2022 年的 56GW 提升至 2030 年的 300GW。

### 推动构建命运共同体　助力非洲加快转型

中国始终是全球绿色转型的重要参与者、贡献者和引领者，在推动自身绿色转型的同时，积极推动国际气候和绿色发展合作，积极向其他国家分享可再生能源发展经验和技术，助力世界各国特别是非洲等发展中国家加快开发利用可再生能源，实现能源低碳转型。2023 年 9 月，中国在非洲气候峰会上宣布，将开发实施应对气候变化的南南合作“非洲光带”项目，利用中国光伏产业优势，打造中非光伏资源利用合作示范带，助力非洲相关国家解决用电困难问题，助力实现绿色低碳发展。

### 非洲经济增长韧性十足　用能需求不断释放

非洲经济保持了较好的增长势头，过去 20 多年间，非洲大陆平均经济增长率为 3.5%，仅次于亚洲。尽管面临新冠疫情、地缘冲突、全球金融紧缩等多重挑战，非洲经济仍展现出十足的增长韧性和发展活力。随着非洲经济发展步伐加快，城市化进程逐步推进，非洲能源需求将进一步增长。发展可再生能源植根于非盟优先发展领域，非盟《2063 年议程》将提升应对气候变化、实现可持续发展作为主要目标之一。随着非洲各国不断推动可再生能源开发，IRENA 预测，到 2030 年，非洲可以通过使用可再生能源来满足其近 1/4 的能源需求。

### 非洲可再生能源资源丰富　合作空间潜力巨大

非洲大陆幅员辽阔，可再生能源资源丰富，可开发水能、风能和太阳能资源分别约占全球的 10%、32% 和 40%，且现有水电、风电、太阳能发电装机规模仅占全球的不足 3%，开发程度相对较低，可再生能源开发空间广阔。近年来，为满足本地能源电力安全保供和经济发展需要，非洲主要国家纷纷在可再生能源发展所需资金、项目等方面积极寻求国际社会支持，同时中国面对以美国为首的西方国家的打压和国际绿色博弈竞争的加剧，也在积极推动可再生能源产业出海，中非双方合作潜力巨大。

### 中国产业技术持续发展　互补优势不断扩大

中国经过多年发展，在风电、光伏发电等可再生能源领域积累了丰富的技术经验，配合较为完整的产业链和产能布局，具有一定的成本和技术优势。中国拥有全球领先的太阳能电池板、电动汽车等新能源产品制造能力，已成为全球锂电池及固态电池领域主要技术来源国之一。随着中非贸易结构持续优化，中国对非洲出口技术含量显著提高，机电产品、高新技术产品对非洲出口额占比超过 50%。借由南南合作、“一带一路”倡议等契机，中国与非洲在可再生能源领域的合作能够极大地发挥双方优势，达成共赢。

### 中国积极推动技术创新　可再生能源持续降本增效

随着中国不断推动光伏发电和风电领域技术进步，如风电机组国产化、大型化进程持续加快，光伏转换效率快速提升，有效降低了有关项目的平均度电成本，为全球可再生能源的普及和应用提供了高品质低成本的可再生能源产品和服务。中国在可再生能源产业蓬勃发展的过程中，将进一步与非洲分享价格合理的风电、光伏发电等绿色能源产品，通过更大范围的开展可再生能源项目合作共建，将非洲巨大的可再生能源资源潜力转化为实实在在的经济增长，提高非洲各国可再生能源的经济性和可及性。

**中非关系处于最好时期　可再生能源合作正当其时**

中非关系正处于历史最好时期，加强同非洲国家的团结合作是中国对外政策的重要基石，也是中国长期坚定的战略选择。几乎所有同中国建交的非洲国家都已与中国签署共建“一带一路”合作文件，非洲已成为参与共建“一带一路”的重要力量。中国支持非盟《2063 年议程》及其旗舰项目的落实，积极参与非洲基础设施发展计划（PIDA）等计划落实。中非双方共同探索可再生能源领域的合作机会，推动贸易和投资的便利化，为可再生能源市场的深入合作创造有利条件，进一步推动中非合作提质升级。

### 4.1.2　合作挑战

**国际形势变化增加合作不确定性**

近年来，全球能源行业跌宕起伏，能源市场供需矛盾不容忽视，经济发展乏力，融资困难增加，导致非洲能源价格上涨、供应紧缩，债务压力陡增，给中非开展清洁能源合作带来一定压力。基于公平公正、互利共赢的原则参与清洁能源项目开发，规避项目开发建设过程中的政治经济风险，减少社会环境影响，帮助非洲建立以清洁能源为主的现代化能源体系，实现绿色可持续发展，是未来中非清洁能源领域合作面临的重要任务。

**发展清洁能源资金短缺问题凸显**

非洲清洁能源发展潜力巨大，但资金投入不足限制了进一步开发利用。2000—2020 年，非洲大陆平均每年仅获得 30 亿美元可再生能源投资额。根据相关研究，在多方努力下，2023 年非洲可再生能源投资额为 120 亿美元，较 2022 年翻了 2 倍，较 2021 年翻了 4 倍。但根据 CMP 预测，在 2040 年前，仅在发电、储能（包括抽水蓄能和电化学储能）和跨境输电方面，非洲每年需要 720 亿美元，现有投资仍无法满足非洲待开发资源潜力和用电需求。

**配套基础设施与人才资源亟须加强**

发展清洁能源对电网灵活调节能力、技术研发创新、物流设备等配套基础设施方面要求较高，且清洁能源项目在设计、建设和运营维护期需要大量专业人员，但是目前当地清洁能源技术基础配套和人才资源存在不足，配套基础设施限制影响清洁能源产业链供应链稳定和清洁能源供应体系建设。此外，非洲能源行业从业者管理水平和技术能力较为欠缺，人才储备不足在一定程度上会增加项目建设成本，限制项目推进速度和后期维护。

**项目商业可持续性有待提升**

非洲大部分国家工商业体量较小，用户侧电价承受能力较低，为促进电力普及，政府对电力用户长期提供高额补贴，财政压力较大，部分国家存在售电电价和上网电价长期倒挂的问题。此外，非洲国家较多，电力市场机制建设不够完善，且政策形势多变，加上中国企业对非洲各国海外商业法律了解不足，缺乏跟踪和应对经验，导致投资项目电费回收存在较大风险，在一定程度上制约了中非电力项目合作，影响非洲电力行业发展。

## 4.2 携手共绘合作新篇章

### 4.2.1 合作方向

**水电助发展 有序推动水电建设**

非洲拥有丰富的水能资源，水电作为非盟 PIDA 规划的重点发展领域，开发潜力巨大。中国作为目前世界上水电发电量最多的国家，水电规划、设计、施工、装备制造、运行维护均居世界先进水平，具备全产业链合作优势。水电具有促进非洲当地社会经济发展、满足电力市场需求、改善基础设施建设、提升交通运输水平、培育属地建材施工产业发展、提高电力系统灵活可靠性等多维度综合效益。中非双方可以以“绿色发展、平等互利、合作共赢”为原则，立足于非洲的发展需求，探索在水电基地开发、流域可再生能源综合利用等领域的合作。

**风光促转型 推动绿色能源升级**

非洲是全球太阳能资源最为富集的大陆，风能资源也很丰富。中国作为全球风能和光伏开发利用第一大国，2023 年中国并网风电和光伏发电合计装机规模达到 10.5 亿 kW，已形成完整的新能源产业链。风电领域在机组大型化、漂浮式风电等方面处于世界先进水平，太阳能领域中光伏组件产能优势显著。新能源发电具有投资小、建设快、惠民生等特点，中非可再生能源合作可以因地制宜地以集中式和分布式相结合的开发模式，满足不同情景下的负荷发展需求，中国可以协助非洲国家开展新能源资源普查、开发规划、项目建设、消纳研究等工作，积极探索非洲海上风电未来开拓前景。

**电网强结构 提升资源配置能力**

非洲因能源资源与市场需求分布不均，电网作为连接电源与负荷的重要纽带，是推动非洲可再生能源发展和电力资源优化配置不可或缺的重要保障。为实现非洲电力资源优化配置，非盟致力打造高效、经济和可持续的 AfSEM，依托电网互联互通建设构建非洲区域骨干大电网。中国电网保持全球特大型电网最长安全纪录，建成并运行世界上输电能力最强、新能源并网规模最大的电网。在电网合作领域，中非可以在打造区域电网互联互通、国内电网规划建设及改造升级等方面深度合作，为打造泛非电力市场提供坚实基础。

**储能稳系统 丰富电源多元结构**

随着非洲新能源高速发展，电网灵活调节能力亟须增加，储能将是保障非洲在新能源大规模、高比例发展情景下电力系统安全稳定的关键举措之一。中国储能技术多元发展、不断迭代，抽水蓄能电站装机规模位居世界第一，锂离子电池、压缩空气储能、超级电容等储能技术达到国际领先水平。中国与非洲国家可以共同开展储能规划配置研究，探索推进非洲抽水蓄能、新型储能等多元化应用与发展，加强在储能系统集成、电池管理系统、储能电站运营等方面的技术交流与合作，助力非洲提升储能领域的自主建设能力。

**微网惠民生 提高能源电力可及**

截至 2023 年，非洲人口超过 14 亿人，居民生活用电需求日益增加。非盟把提高电力覆盖率、减少无电人口、实现电力可及作为非洲电力发展的重要方向。中国通过实施无电地区电网延伸和可再生能源供电工程建设，于 2015 年全面解决了偏远地区的无电人口用电问题。惠及民生用电是中非可再生能源合作的重要组成部分，在解决非洲无电地区的电力供应问题方面，通过借鉴中国经验，提出适宜非洲发展的方案，构建区域微网型电力系统，并通过“方案－示范－推广”的方式，逐步解决非洲大陆电力可及问题。

### 4.2.2 合作建议

**深化合作机制长线服务　发挥跟踪协调和保障作用**

完善长效服务机制，做好非洲区域和国家顶层设计政策研究，搭好合作平台，吸纳合作资源；根据非洲国家实际发展需要，协助制定合理且能落实的清洁能源发展计划。联合非洲有关专业机构，加强相关政策研究，分享中国及第三方清洁能源发展经验。对接非洲基础设施发展规划，以及非洲重点区域、重点国别发展战略，帮助非洲国家做好清洁能源资源普查和发展规划，促进清洁能源项目开发。

**推动绿色金融模式创新　强化风险精准评估和防控**

加大绿色投融资支持，鼓励多元创新模式，开展对非清洁能源产业的中长期投融资活动，支持能够促进非洲当地电力可及的民生项目。在符合商业原则的前提下，为非洲清洁能源产业规划、项目开发以及运营维护等相关领域提供资金支持。鼓励创新绿色金融模式，在项目信息收集基础上，进一步做好非洲清洁能源绿色项目经济可行性、融资风险等评估研判，促进非洲当地绿色可持续发展。

**打造中非合作示范项目　加强技术交流与人才培养**

通过与非洲国家共同打造清洁能源示范基地，加强与非洲国家先进技术交流、应用和知识分享，积极开展技术研讨、专业论坛、联合研究以及项目孵化等多元化形式的国际合作。多形式探索非洲本土人才培养工作，通过能力建设培训、联合办公，设立研发、生产和运输基地，引进先进装备制造生产线和高端人才等方式，帮助提升非洲当地从业者清洁能源治理能力、专业技术以及劳动技能水平。

**共同加强项目保障　提升项目经济可持续性**

中非双方共同加强项目保障，政府层面签署相应的投资保护协定，规避政治与社会因素对项目可能产生的影响，减轻政策变动等意外情况对项目实施的阻力。同时，依托具体的工程项目开展案例研究，通过规划、设计、建设和运营全过程的开发实证，进行长周期跟踪评价，对非洲清洁能源项目开发的实际建设和运营效果进行深入分析，归纳总结项目的经济和社会效益，提升项目经济可持续性。

### 4.2.3 合作愿景

**把握合作机遇　加强战略对接**

当前，中非关系正处于历史最好时期，双方依托“一带一路”、中非合作论坛等合作机制，逐渐形成了全方位、多层次、立体化的合作框架。中非将基于双方共同制定的《中非合作 2035 年愿景》，推动共建“一带一路”合作倡议同非盟《2063 年议程》及非洲各国发展战略对接，不断凝聚合作共识，确立清洁能源领域中长期合作方向，共同推动制定清洁能源领域合作规划，促进中非清洁能源合作高质量持续发展。

**巩固机制建设　共同应对挑战**

中非在共建“一带一路”倡议和中非合作论坛等合作框架下不断深化清洁能源领域合作，以中国-非盟能源伙伴关系为着力点，巩固中-埃、中-摩、中-南（非）、中-纳、中-阿（尔及利亚）等已有双边合作机制下清洁能源合作，积极拓展与更多非洲国家在清洁能源领域的长效合作机制，促进中非在清洁能源领域资源共享、优势互补、互利共赢，保障投资与贸易便利化，加强项目商业可持续性，共同应对合作挑战，保障中非清洁能源合作行稳致远。

**深化产能合作　实现互利共赢**

“一带一路”倡议提出以来，在相关产能与投资机制的推动下，中非加速产能与投资合作。双方将继续秉持相互尊重、互利共赢的原则，加强在基础设施规划、设计、建设、运营、维护等方面的互利合作，统筹非洲清洁能源发展需求与中国产业链技术优势，持续推动合作研究、项目开发、技术转移和成果转化，支持非洲在清洁能源领域建设完善技术标准体系，提升质量基础设施能力，帮助非洲推动制造业升级，提升产业竞争力，创造更多就业岗位。

**共创合作典范　引领绿色发展**

中非聚焦水电、风电、光伏发电等绿色能源领域合作，促进双方实现可持续发展。立足于非洲清洁能源发展需要，充分利用非洲清洁能源资源优势，统筹考虑开发条件，打造示范项目，包括大中型标志性项目，“小而美”精品项目，与农业、矿业等结合的综合利用项目，在探索可持续发展的商业模式等方面树立合作典范，发挥以点带面的引领作用，为中非清洁能源合作提供借鉴和示范，助力提升非洲国家可持续发展能力，为提高人民生活水平作出积极贡献。

**加强国际合作　促进共同繁荣**

中非从来就是命运共同体，中国人民始终同非洲人民同呼吸、共命运，同心相向、守望相助。作为世界上最大的发展中国家和发展中国家最集中的大陆，中国和非洲比以往任何时候都需要加强团结合作。中国将继续呼吁国际社会共同关注中非清洁能源合作与发展，充分彰显中非友好的合作精神，秉持“真实亲诚”理念，讲求义利相兼、以义为先，加强同非洲国家清洁能源领域务实合作，共筑更加紧密的中非命运共同体，为推动构建人类命运共同体树立典范。

# 中非可再生能源合作大事记（2021—2023 年）

## 2021 年

**1 月 6 日** 中国与刚果民主共和国签署了两国政府关于共同推进“一带一路”建设的谅解备忘录，向外界发出中刚致力于共同发展、共同繁荣的积极信号，为两国在包括可再生能源领域在内的各领域互利合作奠定基础。

**1 月 7 日** 中国与博茨瓦纳签署了两国政府关于共同推进“一带一路”建设的谅解备忘录等合作文件。共建“一带一路”将为两国深化互利合作提供新机遇、拓展新领域、开辟新前景，进一步夯实了两国在可再生能源领域的合作基础。

**8 月中旬** 中国企业承建的赞比亚下凯富峡水电站首批机组并网发电。项目总装机容量750MW，是赞比亚在建最大的单体基础设施项目，同时也是该国最大水电站。项目为当地创造了上万个就业机会，为经济社会发展和民生改善创造了新的有利条件。项目配套建设的中国水电培训学院已在当地开班 6 期，300 多名学员顺利毕业，成为该项目乃至赞比亚国内各工程建设的中坚力量。

**10 月 19 日** 中国国家能源局与非洲联盟委员会签署了《中华人民共和国国家能源局和非洲联盟关于中国-非盟能源伙伴关系的谅解备忘录》。根据谅解备忘录，双方同意建立中国-非盟能源伙伴关系，并成立联合工作组。双方将开展政策和信息交流、能力建设、项目合作以及三方合作等领域的合作。

**11 月 22 日** 中国与几内亚比绍签署了两国政府《关于共同推进丝绸之路经济带和 21 世纪海上丝绸之路建设的谅解备忘录》，为两国在包括可再生能源领域在内的各领域互利合作奠定了基础。

**11 月 26 日** 为介绍新时代中非合作成果，展望未来中非合作前景，中华人民共和国国务院发布《新时代的中非合作》白皮书，全面总结了习近平外交思想在非洲方向的新理念、新实践、新成果，用大量数据和事实诠释了全方位、宽领域、立体式的中非合作格局，系统梳理了中非双方在可再生能源等合作领域取得的丰硕成果。

**11 月 29—30 日** 中非合作论坛第八届部长级会议在塞内加尔首都达喀尔举行，会议以“深化中非伙伴合作，促进可持续发展，共同构建中非命运共同体”为主题，中国和非洲国家就加强合作、实现共同发展进行了深入讨论，达成了一系列重要共识。会议通过《达喀尔宣言》和《达喀尔行动计划（2022—2024）》两项成果文件，明确提出，中方将同非方在中国-非盟能源伙伴关系框架下加强能源领域务实合作，共同提高非洲电气化水平，增加可再生能源比重，逐步解决能源可及性问题，推动双方实现能源可持续发展。

**12 月 13 日** 为落实好中非合作论坛第八届部长级会议精神，根据《中华人民共和国政府与非洲联盟关于共同推进“一带一路”建设的合作规划》相关安排，召开中非盟共建“一带一路”合作工作协调机制第一次会议，中方有关部门和非盟委员会围绕包括能源、产能投资、基础设施在内的相关领域合作进行交流，进一步深化了共识。会议签署了《中华人民共和国国家发展和改革委员会与非洲联盟委员会关于建立共建“一带一路”合作工作协调机制的谅解备忘录》。

## 2022 年

**1 月 5 日** 应厄立特里亚外交部长奥斯曼邀请，中华人民共和国国务委员兼外交部长王毅访问厄立特里亚。访问期间，伊萨亚斯总统、双方外长举行会谈会见，双方同意以厄立特里亚加入共建“一带一路”为契机，深化能源、基础设施等领域的合作。

**1 月 5 日** 摩洛哥同中国签署了《中华人民共和国政府与摩洛哥王国政府关于共同推进“一带一路”建设的合作规划》。摩洛哥是北非地区首个与中国签署共建“一带一路”合作规划的国家。该规划进一步夯实了两国在可再生能源领域的合作基础。

**3 月** 由中国公司承建的尼日利亚宗格鲁水电站首台机组正式并网发电。电站是“一带一路”非洲区域重要项目，装机容量 700MW，兼具发电、防洪、灌溉、养殖、航运等多种功能。该工程建设同时解决了当地超过 4000 人的就业问题，对改善尼日利亚电力紧缺局面、增强电网稳定性和持续供电能力、改善人民生活水平等方面具有重要作用。

**6 月** 由中国企业承包的肯尼亚索西安地热电站投产送电。根据肯尼亚制定的 2030 年愿景，该国将在 2030 年前实现 100% 可再生能源发电，其中地热发电的装机容量将达 1600MW，占全国发电量的 60%。索西安地热电站极具示范意义。

**6 月 6 日** 中国国家能源局委托水电总院配合牵头开展中国-非盟能源伙伴关系筹建和运行有关工作，联系双方有关部门、能源企业和研究机构，在中国-非盟能源伙伴关系框架下开展能力建设、联合研究、项目孵化等工作，切实推进能源领域务实合作。

**6 月 15 日** 中非共和国首座光伏电站——萨卡伊光伏电站并网发电。中非合作论坛北京峰会期间，中国和中非共和国就该电站项目的建设达成共识。电站由中国企业总承包建设，装机容量为 15MW，有效缓解了班吉的用电难问题，促进了当地社会经济发展。

**10 月 24 日** 在中国国家能源局和非盟委员会指导下，水电总院以“大规模光伏电站及光伏 +”为主题，举办了伙伴关系框架下第一期能力建设培训。为非方超过 200 位学员授课，旨在助力非洲学员了解和掌握可再生能源领域相关技术的应用，提升管理能力。

**11 月 2—3 日** 在坦桑尼亚联合共和国总统访华期间，中坦两国签署关于建立全面战略合作伙伴关系的联合声明，为两国在包括可再生能源领域在内的各领域互利合作奠定基础。

**11 月下旬** COP27“中国角”会场成功举办“可再生能源赋能非洲电力可及”主题边会。边会上各有关单位在中国-非盟能源伙伴关系框架下联合发布了《可再生能源推动非洲电力可及埃及倡议》和对非研究成果报告《非洲电力可及现状分析及可再生能源离网案例研究》。

## 2023 年

**3—4 月** 中国国家能源局应邀访问纳米比亚和安哥拉。在纳期间，纳总统根哥布欢迎中方企业积极发挥自身优势，参与纳能源发展和项目合作；在安期间，中安双方就可再生能源发展和务实项目合作进行了深入交流，并主持召开在安能源企业座谈会，了解重点项目合作进展情况。

**4 月 18 日** 在加蓬共和国元首访华期间，中加两国签署关于建立全面战略合作伙伴关系的联合声明，为两国在包括可再生能源领域在内的各领域互利合作奠定基础。

**5 月 26 日** 中国与刚果民主共和国签署关于建立全面战略合作伙伴关系的联合声明，为两国在包括可再生能源领域在内的各领域互利合作奠定基础。

**6 月 19 日** 中国国家能源局局长章建华会见应邀来访的南非电力能源部部长洛西恩佐 · 拉莫豪帕。双方就中南电力、可再生能源、核电等合作事项深入交换了意见。

**6 月 20 日** 中国国家能源局局长章建华与来访的坦桑尼亚能源部部长贾努阿里 · 尤瑟夫 · 马坎巴举行双边会谈。双方就油气、可再生能源及电力合作等议题交换意见。

**7 月 10 日** 中国和几内亚比绍签署了共建“一带一路”谅解备忘录，双方同意将两国关系提升为战略伙伴关系，为两国在包括可再生能源领域在内的各领域互利合作奠定基础。

**7 月 17—21 日** 阿尔及利亚民主人民共和国总统阿卜杜勒马吉德 · 特本来华访问，中阿双方同意在“新阿尔及利亚”愿景和共建“一带一路”倡议框架下，通过深化基础设施、能源等领域合作，巩固伙伴关系，为两国各领域合作开辟更广阔前景。期间，两国能源主管部门签署了《关于可再生能源合作的谅解备忘录》，深化两国可再生能源领域交流。

**8 月 14 日** 尼日利亚最大的水电站宗格鲁水电站项目移交证书（TOC）正式签署，标志着项目完工并移交业主。

**8 月 22—24 日** 金砖国家领导人第十五次会晤在南非举办，五国领导人围绕“金砖与非洲：深化伙伴关系，促进彼此增长，实现可持续发展，加强包容性多边主义”主题，就金砖国家合作及共同关心的重大国际问题深入交换意见，达成广泛共识。

会议期间，中南两国政府签署了《中南关于同意深化“一带一路”合作的意向书》，中国企业与南非电力公司签署战略合作备忘录，以响应南非电力领域多项改革举措，积极参与南非可再生能源项目投资，助力南非电网发展和能源转型，实现互利共赢。

**8 月 22 日** 在中国国家能源局和非盟委员会的指导下，水电总院与非盟驻华代表处联合举办中国-非盟能源伙伴关系框架下首届能源合作项目推介会。会议以“分享非洲能源发展机遇，推动中非项目务实合作”为主题，旨在促进中非能源项目合作，鼓励中国企业积极参与非洲基础设施与发展计划（PIDA）及非洲能源项目投资建设。来自非洲 30 多个国家的 43 位驻华使节出席会议。

**8 月 23 日** 在中国国家能源局和中国驻非盟使团的指导下，水电总院在京组织召开中非可再生能源项目信息库建设启动会议，约 25 家主要金融机构和能源企业的领导和代表参会，围绕项目库工作目标、内容和机制进行深入讨论。后续每年定期进行项目信息填报，及时掌握中非能源合作的现状和发展趋势，助力潜在合作项目孵化落地。

**8 月 31 日至 9 月 3 日** 贝宁共和国总统帕特里斯·塔隆来华访问期间，中贝双方签署关于建立战略伙伴关系的联合声明，将进一步深化各领域友好互利合作，进一步夯实了两国在可再生能源领域的合作基础。

**9 月 4 日** 首届非洲气候峰会在肯尼亚首都内罗毕召开，中国代表在发言中宣布，为落实《中非应对气候变化合作宣言》，中国将开发实施应对气候变化的南南合作“非洲光带”项目，助力非洲国家应对气候变化和绿色低碳发展。

**10 月** 由中国企业勘测设计的刚果民主共和国布桑加水电站项目正式投运并发电，刚果民主共和国总统齐塞克迪出席水电站落成典礼并剪彩。 该项目装机总量为 240MW，水电站四台机组全部投运发电后，平均年发电量预计可达 13.31 亿 kWh，约占刚果民主共和国全国电量的 1/10，对当地经济社会发展起到重要作用。

**12 月 8 日** COP28 在迪拜召开。 会议期间举办了“一带一路”绿色低碳转型合作研讨会，聚焦中非能源创新合作。 会上宣布“中非能源创新合作加速器项目”正式上线。 该项目将在中国-非盟能源伙伴关系框架下，遴选并推广助力非洲能源转型的创新案例和创新技术解决方案。

# 缩略词

| APEC | 亚太经济合作组织 |
|---|---|
| AfSEM | 非洲大陆统一电力系统 |
| AfCFTA | 非洲大陆自由贸易区 |
| AFCONE | 非洲核能委员会 |
| AfDB | 非洲开发银行 |
| AU | 非洲联盟 |
| AUC | 非洲联盟委员会 |
| AUDA - NEPAD | 非洲联盟发展署-非洲新伙伴关系 |
| AFREC | 非洲能源委员会 |
| AGHA | 非洲绿色氢能联盟 |
| BRI | “一带一路”倡议 |
| BESS | 电池储能系统 |
| BRICS | 金砖五国（巴西、俄罗斯、印度、中国和南非） |
| CAPP | 中部非洲电力池 |
| COMELEC | 北部非洲电力池 |
| CMP | 非洲大陆电力系统总体规划 |
| COP28 | 联合国气候变化大会第 28 届缔约方会议 |
| CREEI | 水电水利规划设计总院 |
| DEI | 多元化、公平性和包容性 |
| EAPP | 东部非洲电力池 |
| EPC | 设计-采购-施工一体化（工程总承包模式） |
| EIB | 欧洲投资银行 |
| FDI | 外国直接投资 |

| FOCAC | 中非合作论坛 |
| --- | --- |
| G20 | 二十国集团 |
| GDP | 国内生产总值 |
| GWEC | 全球风能委员会 |
| IAEA | 国际原子能机构 |
| IEA | 国际能源署 |
| IRENA | 国际可再生能源署 |
| IFC | 国际金融公司 |
| ICT | 信息与通信技术 |
| MoU | 谅解备忘录 |
| NEA | 中国国家能源局 |
| PIDA | 非洲基础设施发展计划 |
| SAPP | 南部非洲电力池 |
| SEFA | 世界银行非洲可持续能源基金 |
| TOC | 移交证书 |
| UAE | 阿拉伯联合酋长国 |
| UNDESA | 联合国经济和社会事务部 |
| UNESC | 联合国经济及社会理事会 |
| UNCTAD | 联合国贸易和发展会议 |
| UNECA | 联合国非洲经济委员会 |
| WAPP | 西部非洲电力池 |
| WBG | 世界银行集团 |

# 单　位

| MW | 兆（$1\times 10^{6}$）瓦 | TWh | 太（$1\times 10^{12}$）瓦时 |
|---|---|---|---|
| GW | 吉（$1\times 10^{9}$）瓦 | PWh | 拍（$1\times 10^{15}$）瓦时 |
| kV | 千（$1\times 10^{3}$）伏 | EJ | 艾（$1\times 10^{18}$）焦 |
| kWh | 千（$1\times 10^{3}$）瓦时 | GJ | 吉（$1\times 10^{9}$）焦 |

# 声　明



本报告相关内容、数据及观点仅供参考，不构成投资等决策依据，水电水利规划设计总院和非洲联盟发展署不对因使用本报告内容导致的损失承担任何责任。

如无特别注明，本报告各项中国统计数据不包含香港特别行政区、澳门特别行政区和台湾省的数据。部分数据因四舍五入的原因，存在总计与分项合计不等的情况。

本报告部分数据引自联合国贸易和发展会议（United Nations Conference on Trade and Development）、联合国非洲经济委员会（United Nations Economic Commission for Africa）、联合国经济及社会理事会（United Nations Economic and Social Council）、联合国经济和社会事务部（United Nations Department of Economic and Social Affairs）、国际可再生能源署（International Renewable Energy Agency）、国际能源署（International Energy Agency）、国际金融公司（International Finance Corporation）、全球风能委员会（Global Wind Energy Council）、中华人民共和国外交部、中华人民共和国国家能源局、非洲联盟委员会（African Union Commission）、非洲开发银行（African Development Bank）等单位发布的数据，在此一并致谢！

水电水利规划设计总院
China Renewable Energy Engineering Institute

# China-Africa Renewable Energy Cooperation Report 2023

China Renewable Energy Engineering Institute ( CREEI )
African Union Development Agency ( AUDA-NEPAD )

中国水利水电出版社
China Water & Power Press
· Beijing ·

EDITORIAL
COMMITTEE

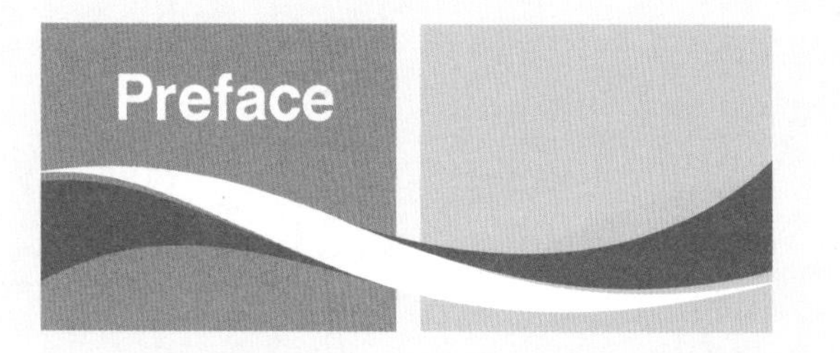


Against the backdrop of the global energy transition, renewable energy development has become an important driving force of economic growth and in pursuit of sustainable development. Africa is one of the continents with the largest renewable energy resources on Earth, occupying an important position in hydroelectric, wind and solar energy resources in the global context. The African Union (AU) has been committed to promoting the continent's energy transition and sustainable development. Rooted in the AU's Agenda 2063 priority projects, developing renewable energy is a long-term vision adopted by African leaders through the AU and is highly consistent with the continent's aspirations for prosperity and inclusive growth. The Program for Infrastructure Development in Africa (PIDA) also takes energy as one of the key development areas. In 2021, the AU Summit adopted the Phase Ⅱ of the PIDA Priority Action Plan (2021 - 2030) of 71 priority projects covering energy, water and other areas, including the Gambia River Basin Organization's hydroelectric power project. Involving more than 40 countries of the AU, the programs are holding a typical significance and scale effect in promoting regional integration and environmental friendliness. The development and utilization of renewable energy are of great significance in satisfying Africa's growing demand for electricity, improving people's well-being and accelerating industrialization. African leaders have repeatedly emphasized Africa's potential and determination in the field of renewable energy on a number of international occasions and called on the international community to support Africa's renewable energy development. In recent years, African countries have quickened the pace of renewable energy development. As a consequence, the continent has significantly increased in the installed capacity of hydropower, wind power and solar power. It is evidenced that share of the installed renewable energy capacity (excluding pumped storage) in Africa's total installed energy capacity grew from 20.2% to 24.6% during the period 2016 - 2023. In 2023, the Africa Climate Summit adopted *The Nairobi Declaration on Climate Change and Call to Action* (*the Nairobi Declaration*), calling on the international community to assist Africa in upgrading its renewable energy power generation capacity and bolstering its installed capacities in renewable energy to 300 GW by 2030.

China has always been a major partner of African countries on the road to renewable energy

development, making great contributions to the development of Africa's infrastructure and renewable energy sectors. As early as 2016, Dr. Ibrahim, the AU Commissioner for Infrastructure and Energy Affairs, pointed out that "about 30% of energy projects in Africa are executed by Chinese enterprises." Most African countries are in a critical period of transformation and upgrading, and China's advanced experience in the field of renewable energy is exactly what African countries need to learn from in their development course. African countries are looking forward to establishing more pragmatic cooperation with China in the field of renewable energy to pursue sustainable economic, social and environmental development, laying a solid foundation for Africa's green future. In the future, the AU remains committed to promoting the development of renewable energy in Africa. The organization is ready to join hands in the international community, especially China and other partners, to jointly push Africa's energy transformation through policy exchanges, strategic alignment, institutional construction and project cooperation. The AU will head towards the beautiful future envisioned in the AU Agenda 2063 at full speed, making every effort to build a new Africa that is peaceful, united, prosperous and self-reliant.

Amb. Rahamtalla M. Osman

Permanent Representative of the African Union to China

Over the past decade, African economies have showcased robust developmental resilience and dynamism, which is expected to continue to be among the fastest-growing regions across the globe. As the AU makes steady advancement in Agenda 2063, the African Continental Free Trade Area ( AfCFTA ) is officially launched, mutual cooperation among subregional organizations has been increasingly strengthened, Africa is emerging as a vital polar of global influence. In parallel with the pursuit of economic growth, the continent will not only have to mainstream climate action into its broader social and economic development activities but also put in place effective adaption measures against the adverse impacts of climate change. effective mitigation of climate change effects will be attained, among other things, by harnessing new technologies, promoting renewable energies, improving efficiency of older energy systems, and changing management practices and consumer behaviour.

Africa is committed to forging a resilient low-carbon economic system. Naturally endowed with about 12% of hydro energy, 32% of wind energy and 40% of solar energy resources on the Earth, Africa boasts a natural advantage in developing renewable energy. With a relatively low electricity penetration rate in Africa, the development and utilization of renewable energy is of great strategic importance for the continent. With these efforts, Africa can not only accommodate the growing electricity demand and enhance people's well-being, but also propel economic growth and achieve sustainable development as a result.

At present, addressing climate change and realizing sustainable development have created a global consensus, and green low-carbon transformation has been quickening worldwide. As a main player in the development of global renewable energy, China has maintained the scale of renewable energy development and utilization ranking the top steadily in the world. By the end of 2023, China's total installed capacity of renewable energy has registered 1,517 GW, accounting for about 40% of the world's total installed capacity of renewable energy, which will provide powerful support for the green transformation of the global energy industry. African countries and China have continuously intensified renewable energy cooperation, effectively promoted green and low-carbon transformation and sustainable development in African regions while guaranteeing the security of energy supply. All

these endeavors have yielded an array of cooperation achievements, which has injected a new impetus into bolstering China-Africa strategic cooperation and achieving sustainable development, demonstrating vivid cooperation dynamics with complementary advantages, mutual benefit and win-win results.

In order to effectively bolster China-Africa energy cooperation, the China National Energy Administration of the People's Republic of China (NEA) and the African Union Commission (AUC) established the China-AU Energy Partnership in October 2021. Commissioned by NEA, the China Renewable Energy Engineering Institute (CREEI) has taken the lead in facilitating the establishment and operation of the China-AU Energy Partnership and fostering the full exchange of policies, strategic planning and project information related to the energy field in China and Africa under the framework of the China-AU Energy Partnership. In doing so, it is useful to help elevate the management capacity and professional and technical level of the relevant personnel of the governmental departments and energy enterprises native to African countries, thereby realizing the multiple goals of sustainable energy development. The year 2024 is coinciding with the summit of the 9th Forum on China-Africa Cooperation (FOCAC) held in Beijing, and on this occasion, CREEI, in preparation with the African Union Development Agency (AUDA-NEPAD) prepares the *China-Africa Renewable Energy Cooperation Report*, of which systematically sorts out the fruitful achievements of China-Africa renewable energy cooperation, and puts forward recommendations and aspirations for promoting the high-quality development of China-Africa renewable energy cooperation. We are in earnest expecting that the Report can provide useful references for the two sides of China-Africa to carry out cooperation in the field of renewable energy.

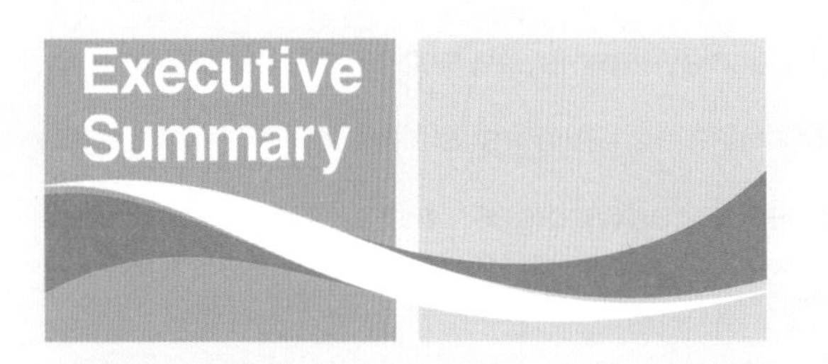


Africa has an abundance in renewable energy resources with a low degree of development, showing huge development potential. Africa has gained strong momentum in its economy in recent years, and with increasing demands for energy supply, African countries have taken the exploits of renewable energy as an important approach to drive energy transformation while meeting the growing electricity demand, as highlighted by the United Nations Economic Commission for Africa ( UNECA ) . Africa and China have set up a solid foundation for cooperation in the renewable energy sector, and Chinese-funded enterprises have built hundreds of hydropower, wind energy, solar energy and other renewable energy projects in Africa, helping African countries in alleviating energy shortages and promoting green development. This Report combs through the development of renewable energy in Africa and China-Africa renewable energy cooperation, summarizes the opportunities and challenges faced by China-Africa cooperation in renewable energy, puts forward cooperation proposals and looks forward to cooperation visions, of which we cherish a desire to provide guidance and practical references for renewable energy cooperation and sustainable development in both sides. The main contents are summarized as follows:

**1. Africa's economy continues to show a strong momentum and unleash the energy demand**

The African economy has shown robust development resilience and dynamism in recent years. from 2019 to 2023, in the face of multiple challenges such as COVID – 2019, geopolitical conflicts and global financial crunch, Africa's real GDP growth rate averaged 2. 1%, in line with the global average growth rate. The United Nations Economic and Social Council ( UNESC ) forecasts that Africa's economy will expand by 3. 5% in 2024, an increase only second to Asia globally. Steady economic growth will contribute to an uninterrupted release of energy demand in Africa. According to the forecast of the African Continental Power System Master Plan ( CMP ), Africa's electricity consumption will arrive at 3,842 TWh by 2040, with a maximum load of 634 GW. As African countries remain committed to the development of renewable energy, the International Renewable Energy Agency ( IRENA ) forecasts that Africa is capable of meeting nearly 1/4 of its energy demand through the exploitation of renewable energy.

**2. African countries are accelerating the development of the renewable energy and effectively promoting the transform of the renewable energy**

Although Africa has energy consumption and production yet dominated by fossil energy and per capita electricity consumption is relatively small. But in the meantime, thanks to its abundant renewable

energy resources, African countries are accelerating energy transformation and promoting the exploitation and utilization of renewable energy. In 2023, the total installed energy capacity in Africa was registered at 252.8 GW, of which the combined installed renewable energy capacity (excluding pumped storage) hit 62.1 GW, accounting for some 24.6% of the total installed capacity. In recent years, Africa has witnessed an increase in installed renewable energy (excluding pumped storage) power generation is significantly higher than that of installed fossil energy power generation. Its growth rate of total installed renewable energy (excluding pumped storage) capacity arrived up to 23.2% during the period 2019 - 2023, which is far exceeding the figure of installed fossil energy capacity (6.4%). According to the CMP forecast, the total size of installed power supply in Africa will reach 1,200 GW, while the installed renewable energy capacity will achieve 750 GW, accounting for 62.5% of the total installed capacity. As a consequence, Africa's energy mix will have a significant opportunity to transition towards a cleaner and more sustainable energy system.

### 3. Africa and China are building a perfect mechanism and deepening the ongoing project cooperation

Since the founding of the People's Republic of China (PRC), China and Africa have been in the same boat and move forward hand in hand for more than seven decades. China has been committed to continuously consolidating mutual political trust between China and Africa, deepening practical cooperation in multiple fields, and providing assistance to Africa's peace and development to the best of its ability, so that China's cooperation with Africa has always been at the forefront of international cooperation with Africa. China has always committed to strengthening the Belt and Road Initiative (BRI) energy partnership and the global renewable energy partnership and proactively contributed to global energy governance through the United Nations, IRENA, G20, BRICS and other cooperation platforms. In line with the Africa Renewable Energy Initiative (AREI) and under the guidance of the BRI and the FOCAC and driven by multi-bilateral cooperation mechanisms, Africa and China have made continuous innovations renewable energy technology. Their project cooperation has brought to fruition, and cooperation mode has evolved from the early days of main foreign aid and engineering contracting towards the drive for integration of investment, construction and operation. Over the past decades, Chinese-funded enterprises have actively responded to the "going global" strategy and built many renewable energy and green development projects in Africa, which cover hydropower, wind energy, solar energy, bioenergy and other fields, yielding a series of cooperation achievements. These projects have more than just focused on economic benefits but also stressed sustainable and inclusive social and environmental resilience, which won high praise from local governments and people and facilitated Africa to follow the road of green and sustainable development.

### 4. Seizing the opportunities for mutual benefit and win-win results and addressing challenges to seek common development

Against the backdrop of global efforts to develop renewable energy, China-Africa renewable energy cooperation is ushering in new development opportunities. Africa's immediate concern is to develop renewable

energy to meet the growing energy demand and China has made great strides in the technology deployed in the renewable energy sector, so both sides have expanding complementary advantages and carry out the renewable energy cooperation well timed. However, bilateral cooperation is challenged by ever-changing international landscapes, Africa's capital shortage, inadequate infrastructure, human resources scarcity and other issues. It is recommended that Africa and China strengthen policy exchange, deepen institutional cooperation, in order to share technology exchanges and talent cultivation program, along with domesticating know how and relevant industries to promote backward and forward linkages with domestic economies. Furthermore, China also should innovate green financial models and enhance project economic sustainability to build cooperation demonstration projects. With mutual support and friendly cooperation with African countries, Africa and China can work together to address risks and challenges.

**5. Strengthening practical cooperation and jointly writing a new chapter in China-Africa renewable energy cooperation**

At the 8th Ministerial Conference of the FOCAC, China announced that it would work closely with African countries to jointly implement the "Nine Programs" such as the Green Development Project, which create new opportunities for the inclusive and sustainable economic and social development of African countries, further unleash the potential of China-Africa renewable energy cooperation as a driving force for the envisaged development process, To further promote the high-quality development of China-Africa renewable energy cooperation, the two sides should strengthen mechanism construction. We need to continue to play the strategic leading role of the FOCAC, intensify policy exchanges and strategic docking, promote the sharing of knowledge and experience, and localize technology and related industries, so as to promote upstream and downstream linkages in the economies of African countries. In addition, the innovation of green financial model is also the key to promote cooperation. China can work with African countries to explore green financial products adopted to the local market and provide financial support for renewable energy projects. In addition, China and Africa should strengthen talent cultivation and technical support, and through technical training and other measures, support African countries to enhance professional capacity in the field of renewable energy. In doing so, we improve the economic sustainability of projects and build bridges for knowledge sharing and technological innovation. With these endeavors, China-Africa renewable energy cooperation will become more in-depth and extensive, which not only promotes the common development of the two sides, but also makes a great contribution to the global energy transition and the fight against climate change. China and Africa will continue to move forward hand in hand on the path of modernization, constantly upgrade the level and hierarchy of renewable energy cooperation, so as to build a high-level China-Africa community of a shared future together.

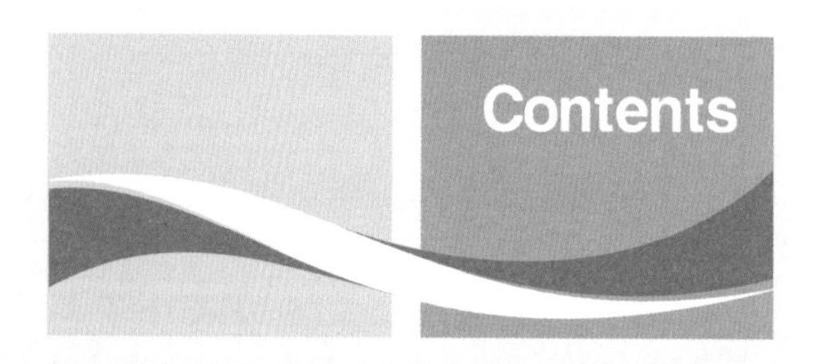
Contents

# 1

# Africa Overview

Situated in the southwestern part of the Eastern Hemisphere and across the equator, Africa is bordering on the east of the Indian Ocean and west of the Atlantic Ocean, across the north of the Mediterranean Sea and Europe. Its northeast to Asia is separated by the Red Sea, with the Suez Canal as the land boundary. The continent covers an area of about 30.37 million square kilometers, accounting for about 20.4% of the total land area of the Earth and shaping the world's second-largest continent (see Figure 1.1). The African continent is shaped wide in the north and narrow in the south, with an average elevation of 750 meters, which is one of the continents with the highest average elevation in the world. It shares a total length of 30,500 kilometers of coastline. Characterized by hot temperatures, low rainfall, and dryness, most of areas are an average annual temperature of more than 20 degrees Celsius, and some areas are scorching all year round, which is hailed as the "Tropical Continent".

Figure 1.1 Africa geographic location map

Africa is the continent with the highest number of sovereign states. For analytical purposes, the United Nations divides Africa into five major regions based on geographical location, taking into account natural conditions and socio-economic characteristics. These regions are North Africa, West Africa, Central Africa, East Africa, and Southern Africa (see Figure 1.2).

Figure 1. 2 Map of geographical regions of Africa (Data source: UNDESA geography program)

## 1. 1 Africa's Demographic Dividend Will Continue to be Released

As of 2023, the African continent had a population of more than 1. 4 billion, about one-sixth of the world's total population and the second largest in the world. Africa has the highest rate of population growth, and according to United Nations Department of Economic and Social Affairs ( UNDESA ) 's Demographic Statistics Division projections, Africa's population will reach 1. 7 billion by 2030 and exceed 2 billion by 2040 ( see Figure 1. 3 ) .

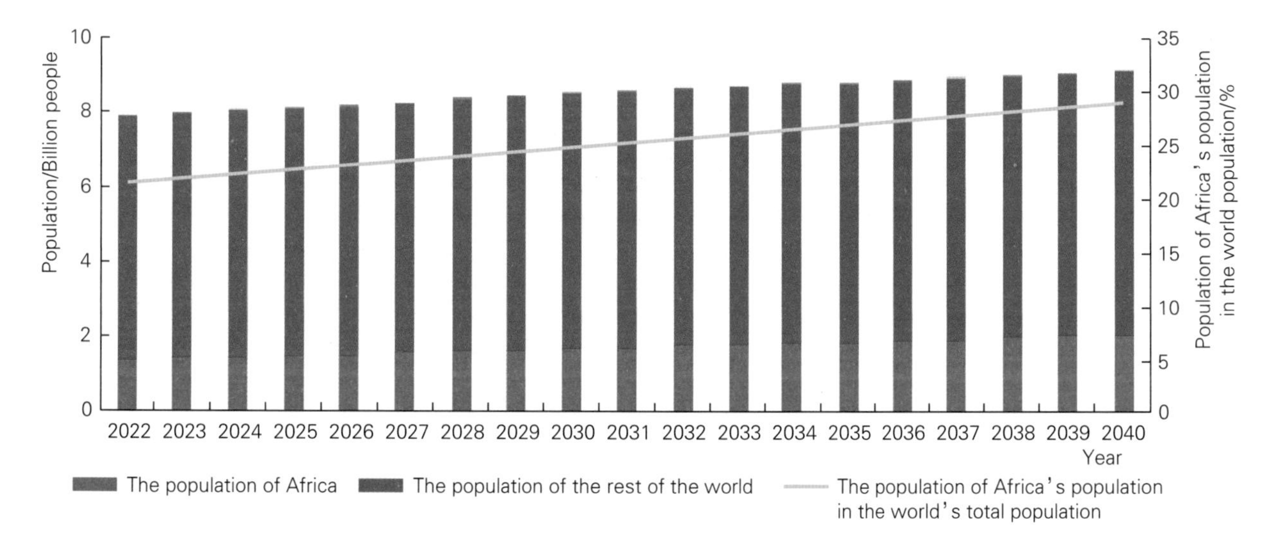


Figure 1. 3 Africa's population projections (Data source: UNDESA demographics division)

According to the World Population Prospects 2024 report, Africa's labor force ( 15 – 64 years old ) reached 844 million people by 2023, more than half of Africa's total population, of which the youth population ( 15 – 24 years old ) was about 289 million, accounting for about 1/5 of the total population ( see Figure 1. 4 ) . Africa's working-age ( aged 20 – 64 years ) population is projected to increase to 1. 6 billion by 2050, accounting for nearly a quarter of the global working-age population, while the youth population is expected to increase to 427 million, implying that there is 1/3 young people globally from Africa. Africa's huge young population signals an abundance of labor pool and huge consumer market potential in the decades ahead. This demographic advantage will continue to unleash a demographic dividend, injecting a steady stream of impetus and vitality into Africa's economic growth.

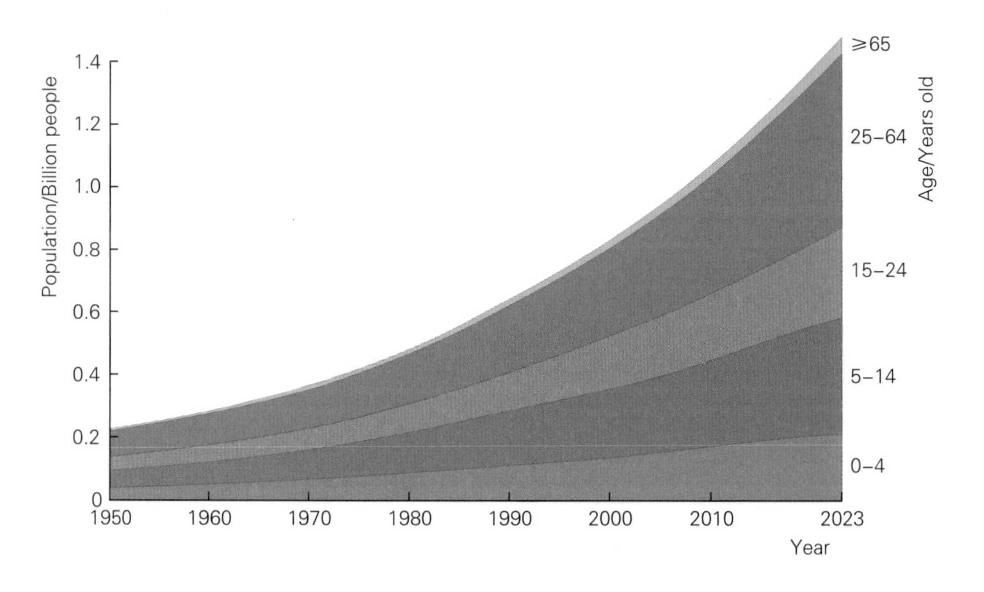


Figure 1. 4 Demographics in Africa (Data source: UN *World Population Prospects 2024*)

## 1. 2 Africa's Economy is Expected to Grow Rapidly in the Long Run

Africa's GDP aggregates reach $ 2. 75 trillion in 2023, with a GDP per capita of $ 1,987. 5, which is about 15% of the global GDP per capita. In the face of multiple challenges such as COVID – 19, geopolitical conflicts, frequent extreme weather, and global financial crunch, Africa's economy has maintained resilient, and according to statistics, that Africa's real GDP would grow at an average rate of 2. 1% during the period 2019 – 2023, which remains roughly unchanged from the global average growth rate ( see Figure 1. 5 ) .

In recent years, as a result of growing intensified global geopolitical tensions, countries heighten concerns about the security of their own energy and mineral supplies, increasing competition for crucial resources. Relying on abundant energy and mineral resources reserves, Africa has accelerated the growth in exports in related fields and foreign investment in recent years, and some African countries have put into additional investment in energy, transportation and other infrastructure construction, leading to continuous contribution of investment and exports to Africa's economic growth during the period 2019 – 2023.

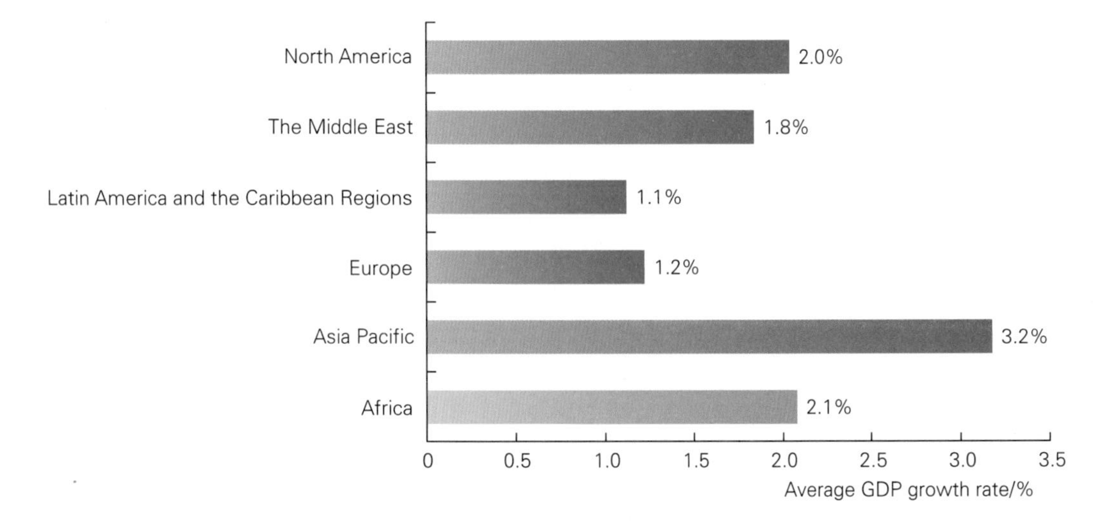


Figure 1. 5 Average global GDP growth rate % of major regions from 2019 to 2023
(Data source: The Economist Intelligence Unit)

In terms of industry structure, the services sector plays a crucial role in the African economy, accounting for more than half of Africa's GDP ( see Figure 1. 6 ), which is primarily contributed by the recovery of tourism in recent years and the high-speed development of the financial sector and information technology. The extractive sector gains strong momentum as well but is limited by the local manufacturing capacity and mostly raw material exports, therefore, upgrading the local manufacturing capacity and increasing the value-added of products has been among the paramount approaches for Africa's economic development. The manufacturing sector has a relatively weak foundation, making up only slightly more than 10% of the total. Africa entails to further promote economic diversification, especially to strengthen the development of manufacturing and other industries in order to achieve balanced economic growth. It is capable of making the best of its advantages in services and natural resources and engaging in international cooperation with other countries to introduce technology and capital to accelerate the upgrading and development of local industries.

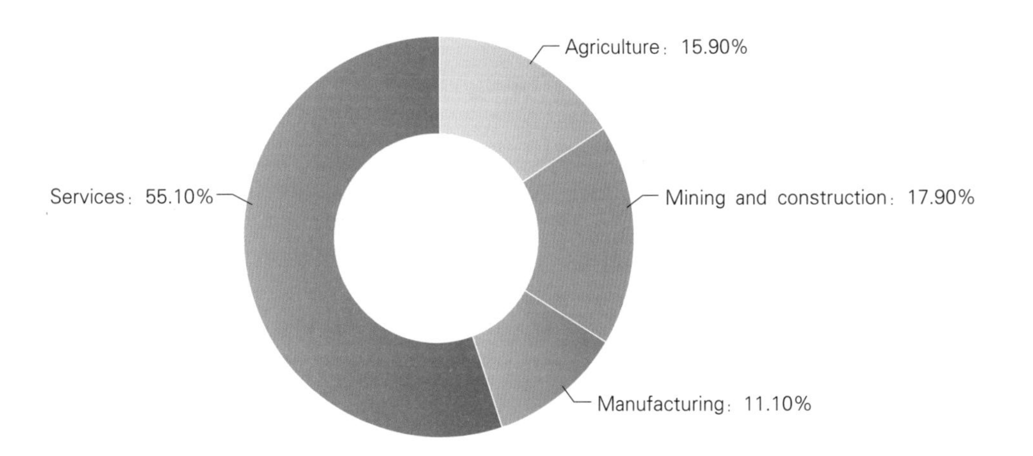


Figure 1. 6 Africa's industrial structure
(Data source: The Economist Intelligence Unit)

The world's largest free trade area is taking shape as the first and second phases of the African Continental Free Trade Agreement negotiations are basically completed. By 2030, AfCFTA would eliminate tariffs on more than 90% of goods and the share of intraregional trade is promised to increase relatively quickly. According to World Bank's projections, by 2035, full trade liberalization will raise Africa's real income by 7%, expand total exports by 29%, increase the share of intraregional trade from the current 13% to 20% (see Figure 1.7), and increase economic output by a projected \$ 212 billion. Regional integration will reduce intra-African trade barriers and promote the free flow and optimal allocation of economic resources in Africa. It also can facilitate countries to participate in market competition and division of labor, fully leverage their comparative advantages, and in the long run, is promising to accelerate Africa's industrialization process.

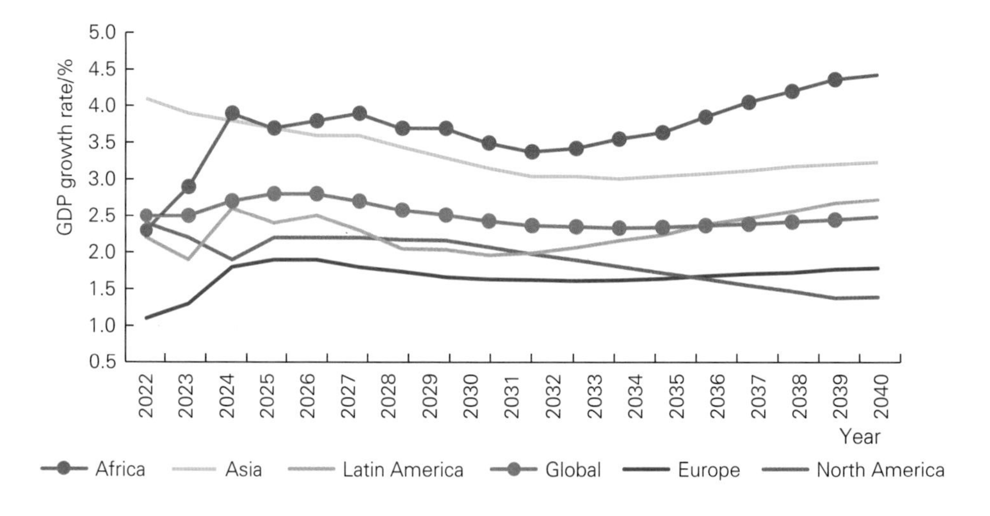


Figure 1.7 Percentage of Africa's Regional domestic and foreign trade [Data source: United Nations Conference on Trade and Development (UNCTAD), World Bank Group (WBG)]

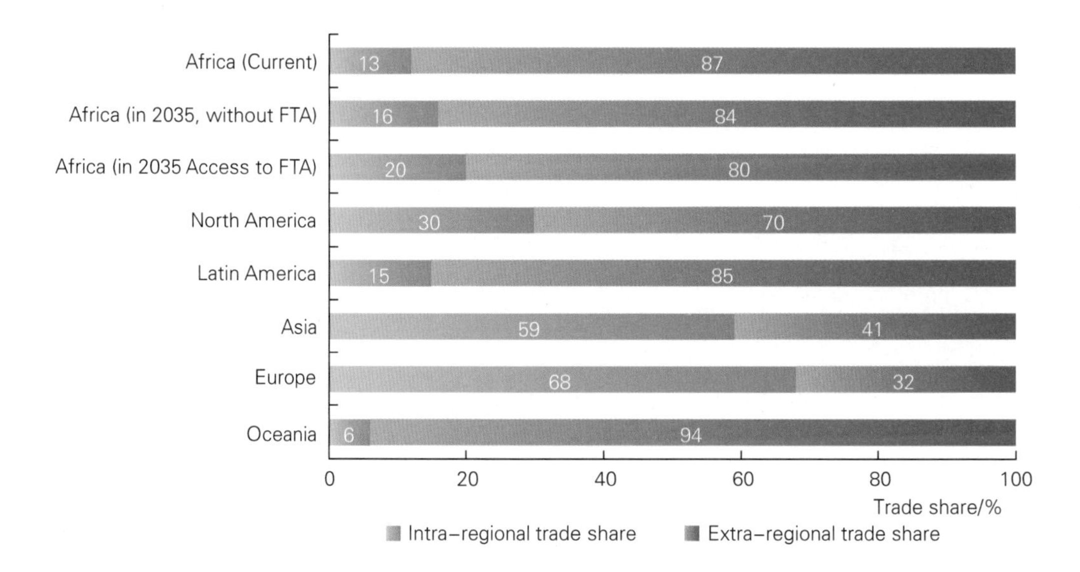


Figure 1.8 Global real GDP growth forecasts by region
(Data source: The Economist Intelligence Unit)

According to the Economist Intelligence Unit ( EIU ) forecast, Africa's economy is promising to rebound between 2024 and 2025, with an average growth rate projected to reach 3.4%. This positive trend is largely driven by increased global demand for Africa's abundant natural resources and a macro environment that is gradually shifting from tight to east monetary policy. Looking ahead, Africa's robust population growth bodes well for continuously unleashing its economic potential. The Africa's average annual GDP growth between 2024 and 2040 will be projected to reach 3.8%, which is 1.3 percentage points higher than the global average at 2.5% ( see Figure 1.8 ), making Africa the fastest-growing region across the globe.

## 1.3 Energy Production and Consumption Levels to be Raised

The African continent has demonstrated positive development momentums in the energy sector. With economic and population growth, expedited industrialization and increased demand for energy, Africa's primary energy production and consumption as a whole shows a steady upward trend from 2011 to 2022. At the same time, Africa is highly adaptive and resilient. Despite the pandemic in 2022 sent the shock to the global economy, which led to a temporary decline in Africa's primary energy production and consumption, it growth resumed in a fast pace. It should be worth noting that Africa's energy production and consumption remain below the world average, however. In 2022, Africa's primary energy production was 35.87 EJ, only accounting for appropriate 5.9% of total global primary energy production; Africa's primary energy consumption was 20.26 EJ, only comprising about 3.4% of total global primary energy consumption, Africa's per capita primary energy consumption falls at 13.9 GJ, about 18% of the world average ( see Figure 1.9 ).

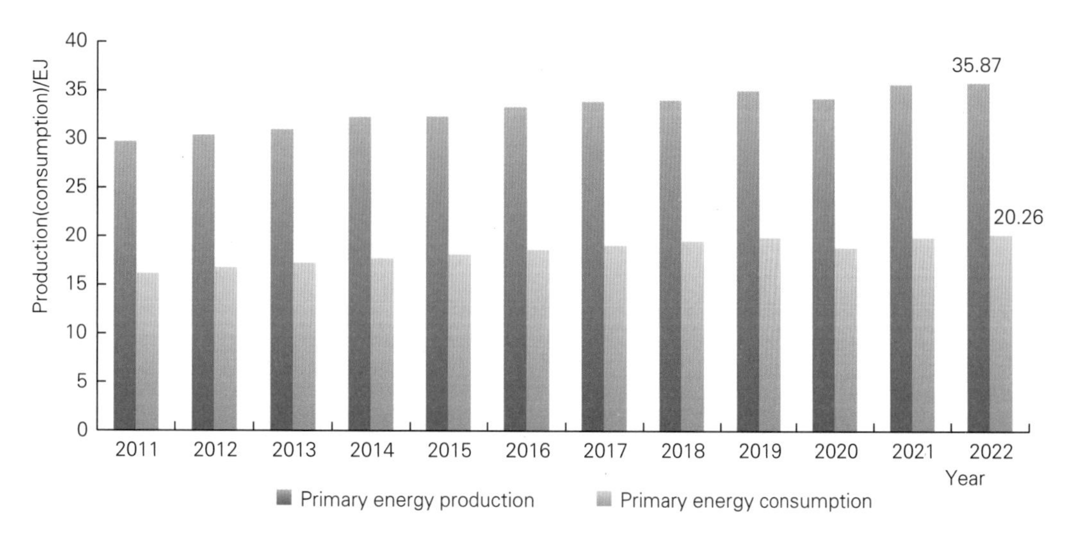


Figure 1.9 Africa's primary energy production and consumption from 2011 to 2022
(Data source: IEA)

Africa's energy mix is dominated by fossil energy and traditional biomass. Oil, natural gas, coal, hydroelectricity and, to a lesser extent, nuclear energy constitute the mainstream of Africa's energy resources. On the energy consumption side, the shares of oil, natural gas and coal consumption in 2022 were

41.41%, 28.87% and 19.60%, respectively, while the share of renewable energy consumption would be relatively low at 9.67% ( see Figure 1.10 ) . On the energy production side, the share of energy production from "other renewable sources" ( mainly solar and wind ) in 2023 was only 1.6%, equivalent to 0.1 billion barrels of oil equivalent. In addition, hydropower accounted for 2.5%, oil for 45.8%, natural gas for 32.3%, and coal for 17.3%. According to the International Energy Agency ( IEA ) study, natural gas contribution will surpass oil by 2028, the share of other renewable energy sources will exceed hydropower by 2033 and coal by 2042 ( see Figure 1.11 ) . As a result, the Africa enjoys massive development potential in renewable energy.

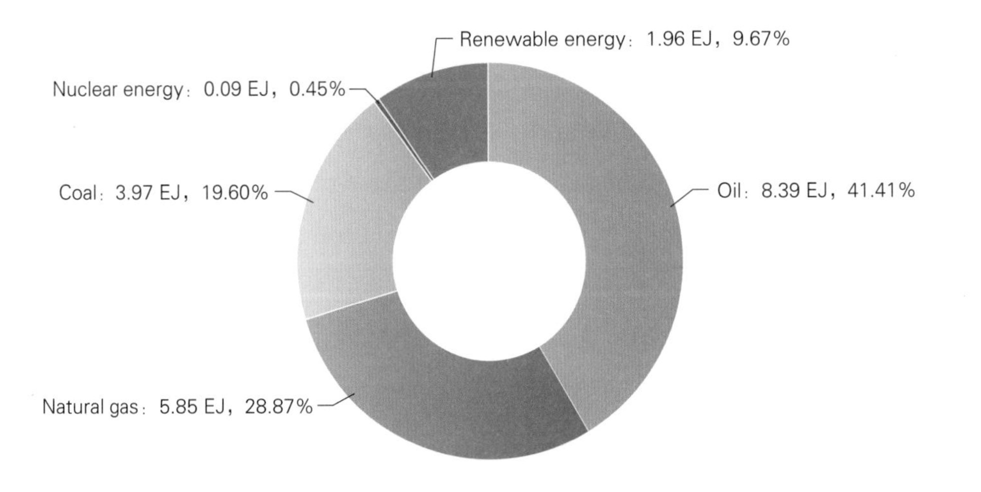


Figure 1.10 African energy sources consumption and structure in 2022
(Data source: IRENA)

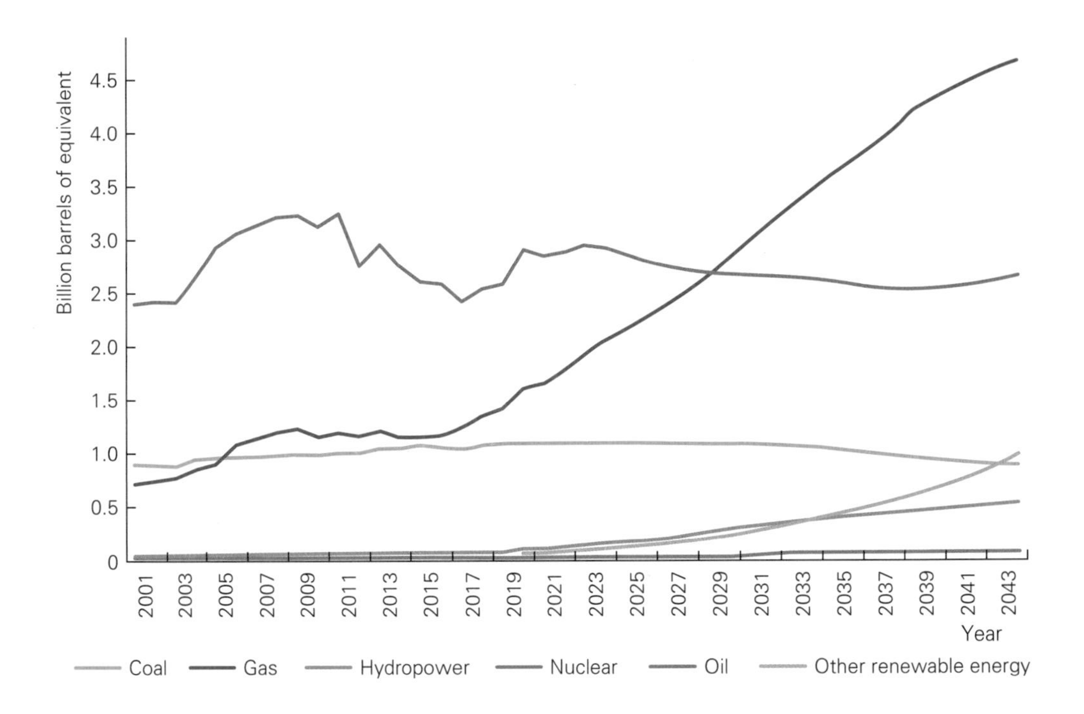


Figure 1.11 The trend of energy production in Africa from 2001 to 2043
(Data source: IEA)

## 1.4 Disparities in Energy Imports and Exports among African Countries

Africa is generally a net exporter of energy, with most coal, unrefined oil and natural gas produced locally and exported to European and Asian markets. According to relevant IEA studies, the value of Africa's energy exports and energy imports will be in balance by 2042 (see Figure 1.12). By 2043, Algeria, Nigeria, Angola and Mozambique will be the top four energy exporters in Africa, with projected exports of $70.2 billion, $32 billion, $24.2 billion and $14.5 billion, respectively. South Sudan ($5.9 billion), the Republic of the Congo ($5.7 billion), Chad ($3.9 billion) and Gabon ($2.7 billion) will also be among the African countries with large energy exports.

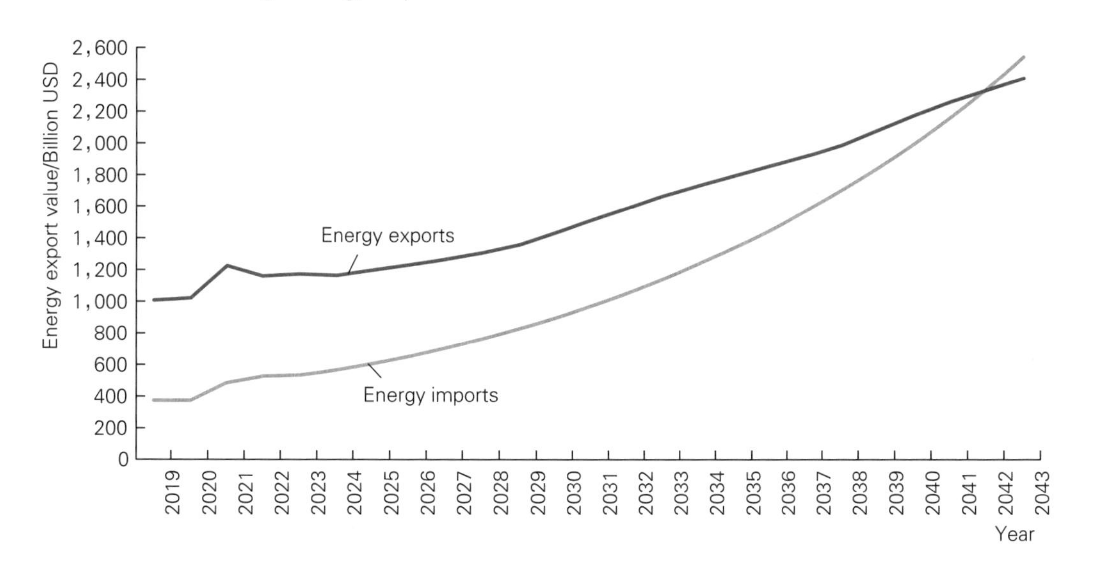


Figure 1.12 2019－2043 Trends in Africa's energy import and export volumes (Data source: IEA)

However, Africa's coal, oil and gas resources are unevenly distributed and concentrated in a few countries, while most other countries remain highly dependent on energy imports, likely threatened by energy security. Currently, energy imports account for more than 50% of the needs in some African countries, such as Senegal (95% of its needs), Morocco, Benin, Mauritania, Mauritius, Eritrea, Tunisia, Uganda, Mali, and Togo (59% of its needs). Burkina Faso, Côte d'Ivoire, Zambia, Ethiopia, Egypt, Niger, Madagascar, Sudan, Rwanda, Libya, Cameroon, and Djibouti are projected to face the same problem with their energy imports by 2043. With the gradual reduction in the cost of renewable energy, most countries in Africa are promising to alleviate their dependence on energy imports and achieve energy self-sufficiency through increased utilization of renewable energy.

## 1.5 Development of Renewable Energy is an Important Way to Promote Sustainable Development in Africa

Africa's electricity installed capacity in 2023 was registered at about 252.8 GW, of which total renewable

energy installed capacity was 62. 1 GW, accounting for about 24. 61% of Africa's electricity installed capacity, including 37. 1 GW of hydropower installed capacity ( excluding pumped storage ), 8. 7 GW of wind energy, 13. 5 GW of solar energy, 1. 9 GW of bioenergy, and 1. 0 GW of geothermal energy; and 1. 9 GW of nuclear energy, accounts for about 0. 75% of Africa's total installed capacity; and 188. 7 GW of fossil energy, accounts for about 74. 64% of Africa's total installed capacity ( see Figure 1. 13 ) .

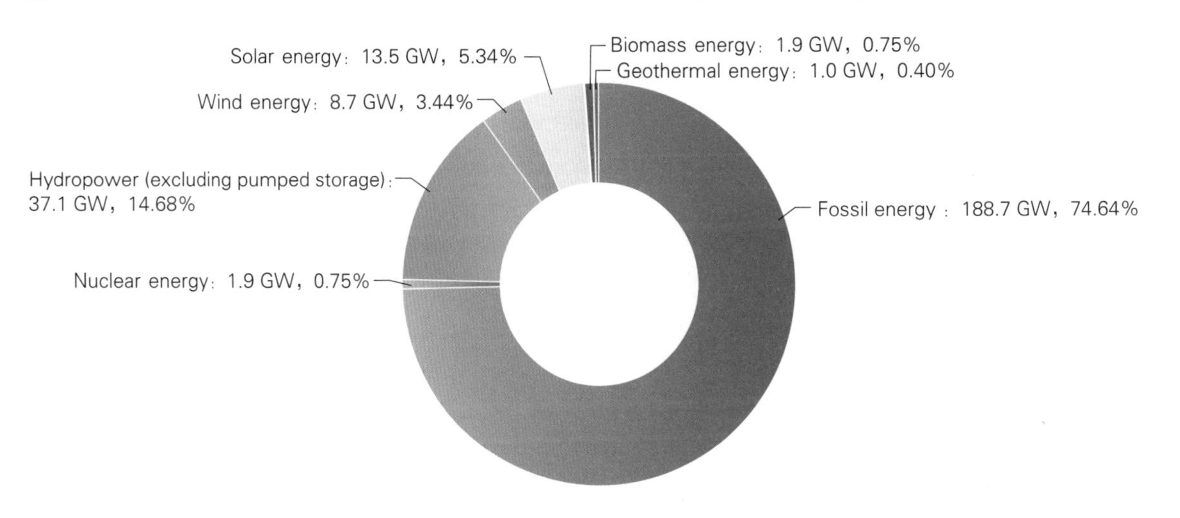


Figure 1. 13 The installed capacity and proportion of various energy sources in Africa by 2023 (Data source: IRENA)

Relevant study shows Africa's total power generation in 2023 arrive in 865 TWh, of which 202 TWh is generated from renewable energy sources, accounting for about 23. 35% of Africa's total power generation capacity, including 159 TWh of hydropower generation capacity, 14 TW of wind power generation capacity, 21 TWh of solar power generation capacity, 3 TWh of biomass power generation capacity, and 5 TWh of geothermal power generation capacity; and nuclear power generation capacity is 13 TWh which accounts for about 1. 5% of Africa's total electricity generation; and fossil energy generation capacity is 650 TWh, which accounts for about 75.15% of Africa's total electricity generation ( see Figure 1. 14 ) .

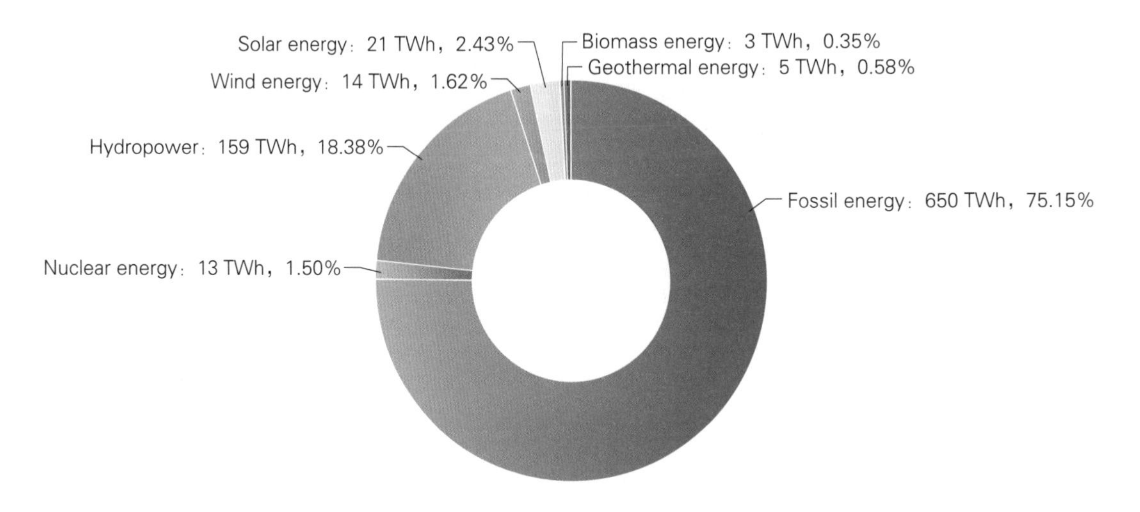


Figure 1. 14 The electricity generation and proportion of various power sources in Africa by 2022 (Data source: CREEI)

Currently, Africa has established five power pool organizations, namely, the East African Power Pool (EAPP), Comité Maghrébin de L'Electricité (COMELEC), the Central African Power Pool (CAPP), the West African Power Pool (WAPP), and the Southern African Power Pool (SAPP), which include 49 countries in Africa. Additionally, several island countries such as Mauritius, Madagascar, and Cape Verde operate independent power grids. Currently, although between member countries within each pool and among pools have set up (or planned) power lines for the exchange of electricity, the problems of electricity unavailability and scarcity in Africa are still grim, and the problems of unbalanced electricity development and inadequate access to power grids are still borne out.

According to the IEA, Africa's per capita electricity consumption in 2023 was about 530 kWh, which is only one-fifth of the world average, while the per capita electricity consumption in sub-Saharan Africa (excluding South Africa) was about 190 kWh. According to the African Development Bank (AfDB) statistics in 2022, Africa's electrification rate is only 40% and more than 640 million people are living in the area of without access to electricity, which is mainly distributed in sub-Saharan Africa. About two thirds of African countries have an electricity access rate of less than 50% (see Figure 1. 15). Inadequate electricity supply in Africa has represented a key bottleneck to its sustainable development, severely curtailing Africa's economic and social development.

Africa is vigorously devoted to renewable energy as one of the key pillars of its future power system. In 2018, the AU proposed to build an Africa Single Electricity System (AfSEM), which aims to enhance electricity accessibility and support sustainable development. In September 2023, the AU approved the CMP, which carries out three circumstances of power development projections with Africa aiming to fully achieve electricity accessibility by 2030 (high growth rate), 2035 (medium growth rate) and 2040 respectively (low growth rate) (see Figure 1. 16). According to the medium growth scenario scheme, it is expected that by 2040, Africa's electricity consumption will reach 3,842 TWh with a maximum load of 634 GW and a total installed capacity of 1,200 GW. Of which the installed capacity of renewable energy will grow from 62. 1 GW in 2023 to 750 GW, and the share of renewable energy in Africa's total installed capacity will increase from 24. 6% in 2023 to 62. 5% and PV and wind power installations are expected to rise to 262 GW and 342 GW respectively (see Figure 1. 17).

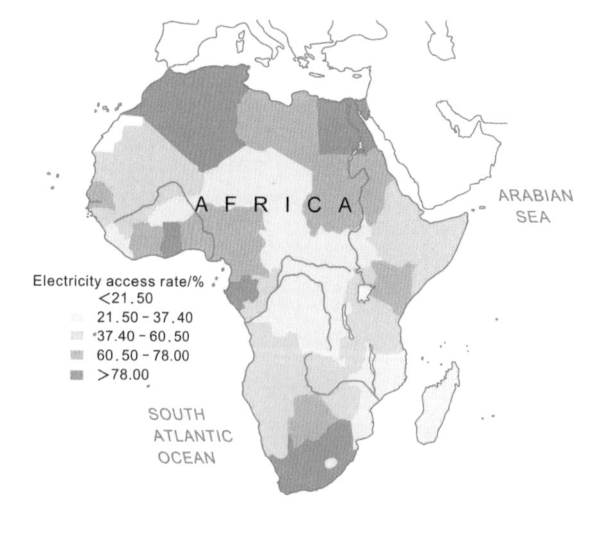


Figure 1. 15 Access to Electricity in Africa by country, 2022 (Data source: World Bank)

In order to address the problem of insufficient power supply in Africa and promote sustainable economic and social development, Africa is vigorously developing renewable energy relying on the characteristics of its energy resources; strengthening the construction of power grid infrastructure in African countries; expanding

the coverage of the power grid and enhancing power supply efficiency through the renovation, upgrading and construction of new transmission and distribution grids; and quickening the promotion of intra-continental and trans-continental electricity interconnections to optimize the allocation of power resources in Africa.

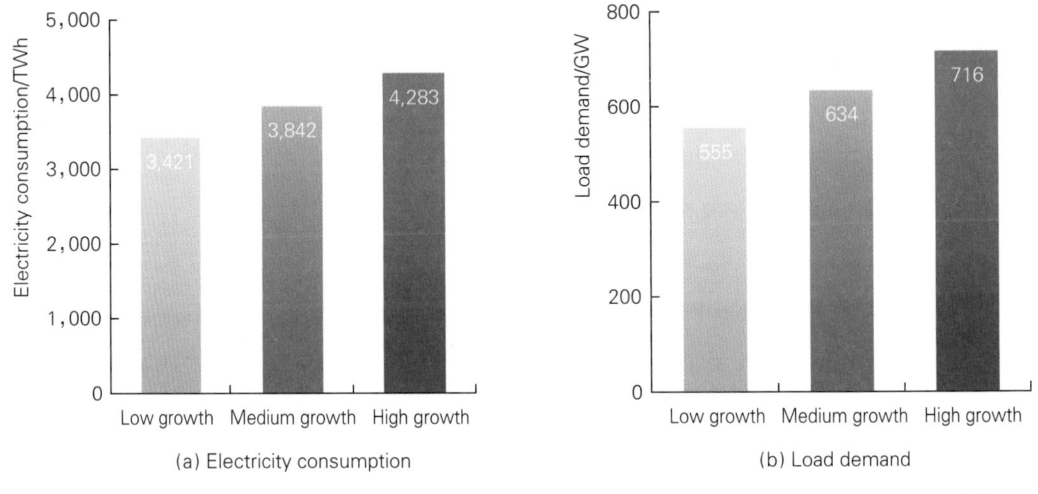


Figure 1.16　Forecasts results in three different circumstances by CMP (Data source: CMP)

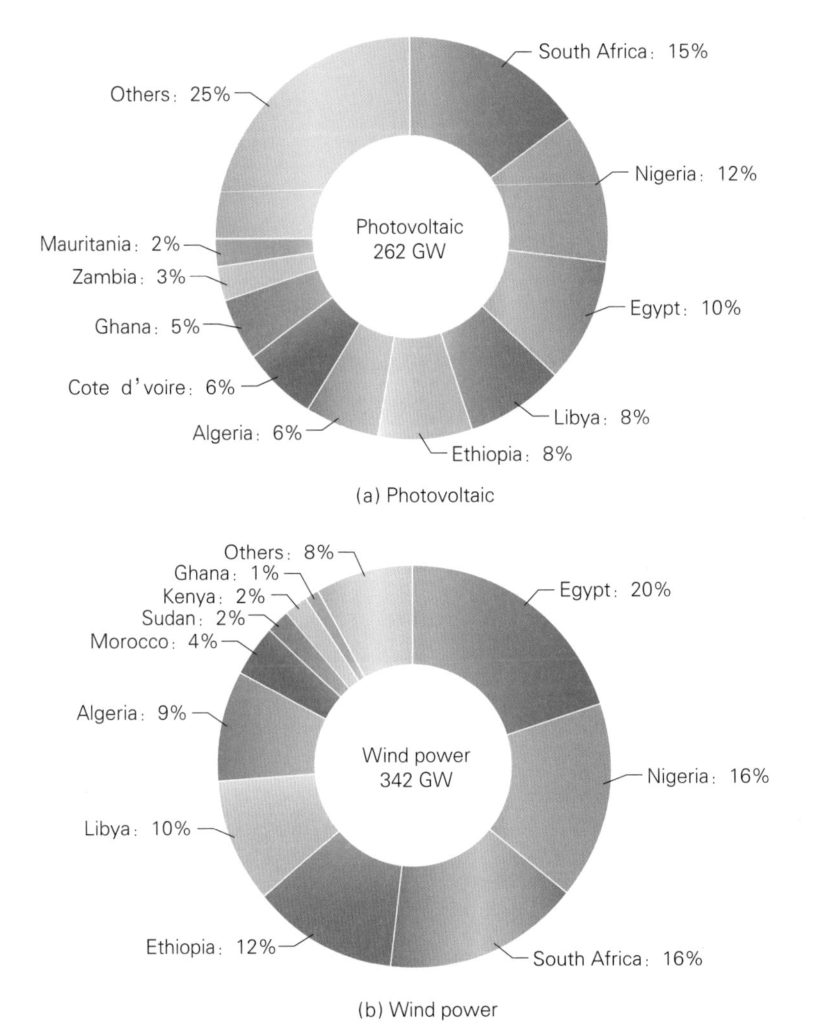


Figure 1.17　Predicted installed PV and wind power distribution in Africa by 2040 (Data source: CMP)

The development of renewable energy is of great significance to the promotion of sustainable development in Africa. The African continent is rich in renewable energy resources, and the development and utilization of these resources is instrumental in increasing Africa's energy self-sufficiency and reducing its dependence on fossil fuels, while at the same time reducing greenhouse gas emissions and combating climate change. In addition, renewable energy projects typically have a low environmental effect, availing for conserving Africa's ecosystems and biodiversity. By investing in renewable energy, African countries can create jobs opportunities, boost local economies and increase the general accessibility of energy, especially in remote areas.

## 1.6 A Multidimensional Analysis of Urban Development and Electricity Consumption in Africa

### 1.6.1 Urbanization Trends in Africa

Africa is experiencing unprecedented urban growth, with projections indicating that by 2050, 60% of its population will live in urban areas. This rapid urbanization, driven by both rural-to-urban migration and natural population increases, has profound implications for infrastructure, particularly electricity consumption. According to the Africa's Urbanisation Dynamics 2020 report, urban populations are set to double in the coming decades, a phenomenon that will have far-reaching impacts on economic development, environmental sustainability, and social well-being.

Currently, 45% of Africa's population resides in urban centers. This trend is placing immense pressure on electricity supply systems, with urban areas being the primary consumers of electricity. Meeting the demands of this growing urban population will require robust investments in infrastructure, energy generation, and distribution networks, especially as the urban electricity consumption is expected to increase significantly in the coming years.

### 1.6.2 Urbanization and Electricity Consumption

Africa's electricity demand is projected to increase fivefold by 2040, rising from 700 TWh in 2022 to 2,368 TWh. Urban centers will contribute substantially to this surge, as electricity consumption per capita in African cities, which averaged 530 kWh in 2023, is expected to rise as access to electricity improves and economic conditions strengthen. Despite these improvements, African urban areas lag significantly behind the global average electricity consumption of 3,000 kWh per capita.

Urban electricity consumption is strongly influenced by several factors, including population growth, economic activities, and industrialization. In the context of African cities such as Lagos, Kinshasa, and Nairobi, meeting the growing electricity demand will require not only grid expansion but also the integration of renewable energy sources and smart infrastructure investments. These efforts are essential to modernize urban electricity

grids and avoid the frequent outages currently observed in many African cities.

### 1. 6. 3 Investment Needs for Electricity Infrastructure

To keep pace with the anticipated rise in electricity consumption, Africa needs to invest between \$40 billion and \$70 billion annually in electricity infrastructure. This investment is crucial for expanding and modernizing the grid, integrating renewable energy, and ensuring that cities can meet the growing demand for electricity. By 2040, urban electricity consumption is expected to increase by over 250%, a challenge that will require governments and development partners to coordinate efforts and mobilize the necessary financial resources.

Moreover, Africa's infrastructure challenges are compounded by the risk of slum expansion and uneven electricity distribution. Cities like Johannesburg and Cairo have experienced frequent electricity outages due to overburdened grids. These examples underscore the need for comprehensive urban planning that includes provisions for sustainable electricity generation and distribution.

### 1. 6. 4 Challenges for the Next Decade

One of the most pressing challenges facing African cities is the issue of electricity access. Despite significant progress in electrification, over 600 million Africans still lack access to reliable electricity. Urbanization is set to exacerbate these challenges as more people migrate to cities in search of better economic opportunities. To meet these needs, African countries must pursue ambitious grid modernization initiatives, promote the use of renewable energy, and adopt innovative solutions such as microgrids, especially in informal settlements where grid expansion may be less feasible.

Additionally, sustainability remains a top priority. Africa is heavily reliant on fossil fuels for electricity generation, but the global shift toward renewable energy presents an opportunity to transition to more sustainable sources of power. Investments in renewable energy not only offer a solution to Africa's growing electricity needs but also align with global efforts to combat climate change and promote sustainable development.

### 1. 6. 5 The Way Forward

The future of urban development and electricity consumption in Africa hinges on the continent's ability to address its infrastructure gaps, modernize its electricity grids, and integrate renewable energy sources into its energy mix. Africa's cities are key drivers of economic growth, and ensuring a reliable supply of electricity is essential for their continued development. Governments, international partners, and the private sector must work together to ensure that the necessary investments are made. By addressing the challenges of electricity access, grid modernization, and sustainability, Africa can harness the potential of its urban transition to foster inclusive and sustainable development for future generations.

## 1.7 Electrification and Digitalization in Africa: A Pathway to Unlocking Socio-Economic Potential

Africa is at a critical juncture in its development trajectory, with rapid urbanization, a growing population, and the promise of digital transformation offering new avenues for socio-economic advancement. However, the continent's potential for digitalization and economic growth is significantly constrained by a persistent challenge: inadequate access to reliable electricity. Digitalization, and by extension the development of a robust digital economy, is inextricably linked to the availability of reliable, affordable, and sustainable electricity. As the continent seeks to integrate itself into the global digital economy, electrification remains a vital prerequisite.

### 1.7.1 Electricity and the Foundation for Digitalization

Electricity access is fundamental to any modern economy, yet over 600 million Africans still lack reliable electricity, particularly in rural and peri-urban areas. This deficit poses a major obstacle to Africa's digitalization efforts. According to the USAID Digital Policy (2024 – 2034), the ability to build open, inclusive, and secure digital ecosystems is essential for socio-economic development, but these digital ecosystems cannot function without a stable and sufficient energy supply.

As Africa's digital economy holds the potential to contribute up to $ 180 billion to the continent's economy by 2025, solving the electricity challenge becomes more urgent. Data centers, cloud infrastructure, mobile networks, and e-commerce platforms are all energy-intensive and require a stable grid to operate effectively. Furthermore, central to digital economies – the integration of AI technologies, relies on continuous and uninterrupted access to electricity. For instance, AI is projected to double the GDP growth rate of African countries by 2035 if harnessed properly. However, achieving this requires sustained investment in electrification.

### 1.7.2 Electrification as a Catalyst for the Digital Economy

Africa's per capita electricity consumption currently stands at 530 kWh, significantly below the global average of 3,000 kWh. This gap underscores the energy challenges that constrain the continent's digitalization efforts. The growth of digital services, data-driven industries, and other forms of digital economic activity depend heavily on widespread, affordable, and reliable electricity. Without a reliable power supply, the expansion of digital infrastructure like data centers, mobile networks, and cloud computing platforms will remain hindered. Moreover, as African economies become more reliant on AI, the demand for power will only increase.

For instance, the African Union's Continental Artificial Intelligence Strategy emphasizes that AI will be a key driver of development in sectors such as agriculture, healthcare, education, and public service delivery. But this potential will only be realized with the necessary infrastructure in place, including reliable electricity. AI systems, particularly those that process large datasets or run machine learning algorithms, are energy-

intensive. Inadequate electricity means that even where AI technologies are available, their application will remain limited.

### 1.7.3 Investment in Renewable Energy

The need for sustainable and reliable electricity across Africa offers an opportunity for investment in renewable energy sources. The African Union's Strategy on Artificial Intelligence highlights the importance of renewable energy in ensuring that Africa's digital transformation aligns with global sustainability goals. Over 60% of new electricity investments on the continent must come from renewable sources, not only to meet rising energy demands but also to support broader environmental sustainability objectives.

Solar energy and microgrids offer promising solutions for rural electrification, which in turn supports the expansion of digital and mobile services in underserved areas. These decentralized energy solutions provide a practical approach to bridging the electricity gap, enabling more communities to participate in the digital economy. Investment in off-grid and mini-grid solutions could help mitigate Africa's electricity challenges, especially in remote areas where extending the traditional grid would be cost-prohibitive.

### 1.7.4 Challenges in Electrification and Digitalization

Despite the significant potential of digital technologies to drive socio-economic transformation in Africa, several challenges remain. First, the issue of affordability-both of electricity and of digital services-remains a major barrier. Many African households and businesses cannot afford consistent access to electricity, let alone the digital services that rely on it. As the AI for Africa report notes, any strategy aimed at fostering digital growth must address the broader issue of infrastructure affordability.

Second, there is the challenge of uneven distribution. Electrification rates vary widely across African countries and even within countries, creating pockets of digital exclusion. This digital divide is especially pronounced in rural areas, where the lack of electricity is compounded by the absence of internet infrastructure. To address this, policymakers need to prioritize investment in both energy and digital infrastructure to ensure that no one is left behind in the digital economy.

### 1.7.5 A Vision for the Future: Digital Growth and Electricity Access

Bridging the electricity gap is not just about providing power; it is about creating an enabling environment for digital innovation and economic growth. Africa's young, tech-savvy population is well-positioned to harness the benefits of digital technologies, but without electricity, their potential remains stifled. Governments, international development agencies, and the private sector must work together to ensure that Africa's electrification goals align with its digitalization ambitions.

Investment in renewable energy and innovative grid solutions must be coupled with efforts to expand internet access, improve digital literacy, and foster an entrepreneurial ecosystem that can take advantage of digital

opportunities. For example, the Continental AI Strategy envisions a future where AI-driven innovations in agriculture, health, and education can transform African societies, but only if the foundational infrastructure, particularly electricity, is in place.

The road to digitalization in Africa is inseparably tied to solving the continent's electricity challenges. Reliable and sustainable energy infrastructure is crucial for powering the digital economy, supporting data-driven industries, and ensuring that technologies such as AI can be fully harnessed to spur socio-economic transformation. Without addressing the current electricity deficits, particularly in rural and underserved regions, Africa risks falling behind in the global digital economy. Strategic investments in renewable energy, grid expansion, and electrification of remote areas will not only bridge the electricity gap but also unlock Africa's potential in the digital era. As the continent embarks on its digital transformation, electrification must remain a priority, underpinning all efforts toward building a resilient and inclusive digital economy.

# 2

# African Renewable Energy Resources and Development Overview

## 2.1 Africa is Rich in Renewable Energy Resources

### 2.1.1 Hydro Energy Resources

Africa boasts an abundance of hydro energy resources and the main rivers are, the Nile River, the world's first long river, Congo River (Zaire River), the world's second-largest water system, as well as the Niger River, the Zambezi River and among others (see Figure 2.1). In theory, African river water energy resources have a reserve of 4 trillion kWh, accounting for about 10% of the world; the technically developable amount of water resources is about 1.75 trillion kWh, about 12% of the world. By 2022, Africa only utilizes 11% of water energy development, empowering huge development potential and development space.

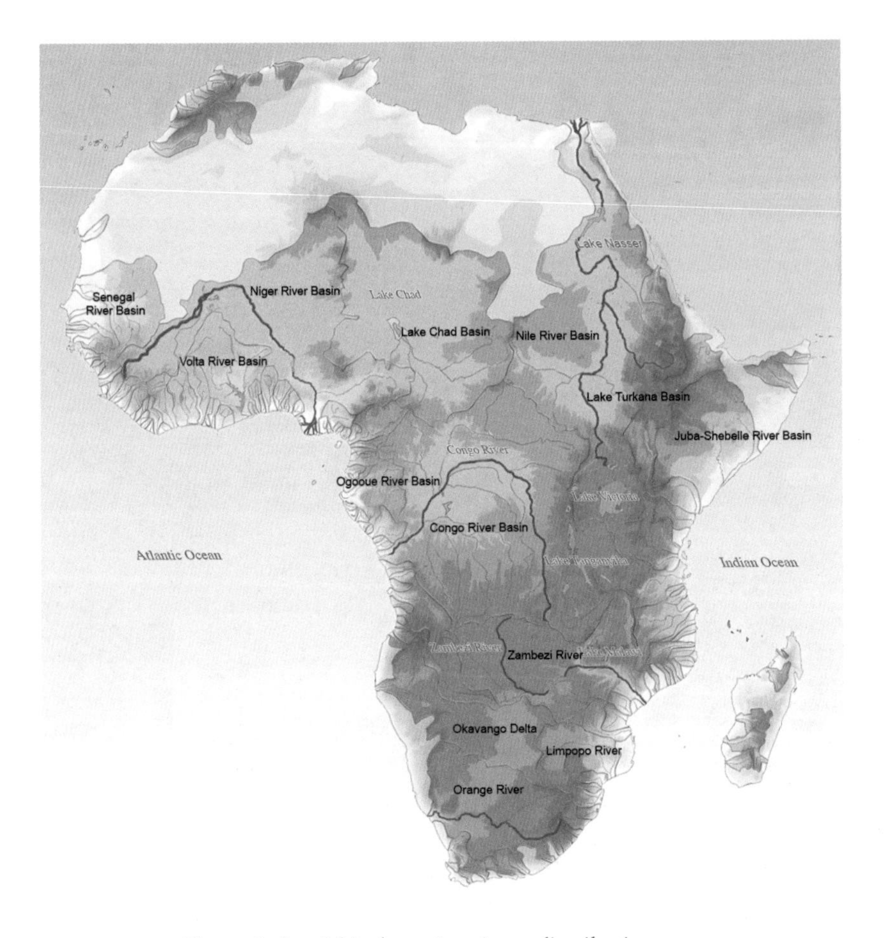


Figure 2.1 Africa's major rivers distribution map

## Nile River

Located in northeastern Africa, the Nile River originates from the Burundi Plateau and flows into the Mediterranean Sea from south to north. Known as the longest river on the Earth, it has a total length of 6,670 km with a catchment area totaling 2.87 million $km^2$, and a multi-year average runoff of 84 billion $m^3$. The two main tributaries of the Nile River are the White Nile and the Blue Nile. The former is 3,700 km in length, originating in the Great Lakes region of central Africa, and flows through Uganda, South Sudan and Sudan; the latter is 1,600 km long, originating in the Ethiopian highlands, and flows through Ethiopia and Sudan.

## Congo River

The Congo River, also known as the Zaire River, is located in the west-central region of Africa and originates in the territory of Zambia, with a total length of about 4,640 km, making it the longest river in west-central Africa and the second longest in Africa. The Congo River runs through Zambia, the Democratic Republic of the Congo, the Republic of the Congo and Angola, and finally converges into the Atlantic Ocean. With a catchment area of 3.7 million $km^2$, the river is famous for an annual runoff of 13,000 $m^3$, and a theoretical reserve of 390 million kW of hydro energy.

## Niger River

The Niger River is located in West Africa north of the equator and originates in the mountains of the Fouta Djallon Plateau in Guinea. With an altitude of about 900 m above the seal level, the river is totaled about 4,200 km in length, listing the third largest river in Africa. The Niger River draws an inverted "U" shape in West Africa, with a catchment area of about 2.09 million $km^2$, running through the countries such as Guinea, Mali, Niger, Benin and Nigeria, and is hailed as the "mother river" of West Africa.

## Zambezi River

The Zambezi River is located in the southeastern region of Africa and originates from the mountains on the northwestern border of Zambia at an altitude of 1,300 m above sea level, much closer to the source of the Congo River. Due to the river's total length of about 2,660 km, the river is Africa's fourth-largest river and the first-largest river in Africa south of the equator. The Zambezi River has a catchment area of about 1.35 million $km^2$, and its main stream travels through six countries, including Angola, Zambia, Namibia, Botswana, Zimbabwe and Mozambique.

### 2.1.2 Wind Energy Resources

Africa's total wind energy resources account for about 32% of the total global wind energy resources. However, the overall distribution is uneven, mainly concentrated in the Sahara Desert and its northern region, the southern coast and the east-central coastal region. Among them, the average wind speed at 100 m height in some areas of Somalia, Mauritania, Chad, Egypt and other countries is above 8 m/s (see Figure 2.2), the wind power density of the Sahara Desert and its northern region, the coast of the Somali Peninsula and the coast of southern Africa can reach more than 400 W/m$^2$ at 100 m height (see Figure 2.3). According to International Finance Corporation (IFC) calculations, Africa has a wind energy potential of 18,000 TWh/year, about 250 times the continent's current electricity demand. The continent's total technical potential capacity of wind energy is about 33,641 GW, of which Northern Africa is about 18,822 GW, Southern Africa is about 891 GW, Eastern Africa is about 2,133 GW, Western Africa is about 9,144 GW and Central Africa is about 2,651 GW (see Table 2.1).

**Table 2.1 Technical potential capacity of wind energy in Africa (Data source: IFC)**

| Regions | Technical potential capacity of wind energy/GW | Regions | Technical potential capacity of wind energy/GW |
|---|---|---|---|
| Northern Africa | 18,822 | Eastern Africa | 2,133 |
| Western Africa | 9,144 | Southern Africa | 891 |
| Central Africa | 2,651 | Total | 33,641 |

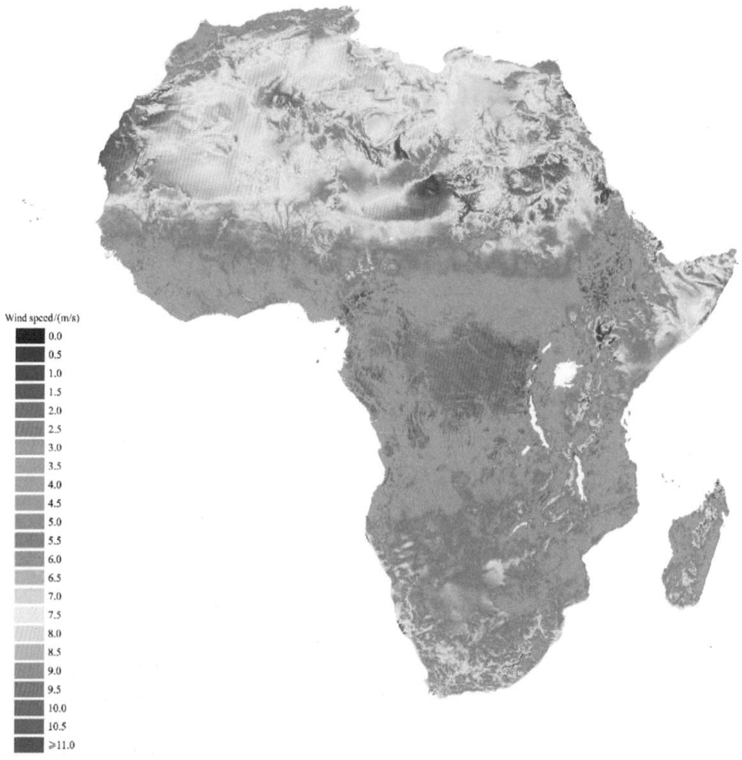

Wind speed/(m/s)
0.0
0.5
1.0
1.5
2.0
2.5
3.0
3.5
4.0
4.5
5.0
5.5
6.0
6.5
7.0
7.5
8.0
8.5
9.0
9.5
10.0
10.5
≥11.0

Figure 2.2 Africa wind speed distribution map at 100 m height
(Data source: Global Wind Atlas)

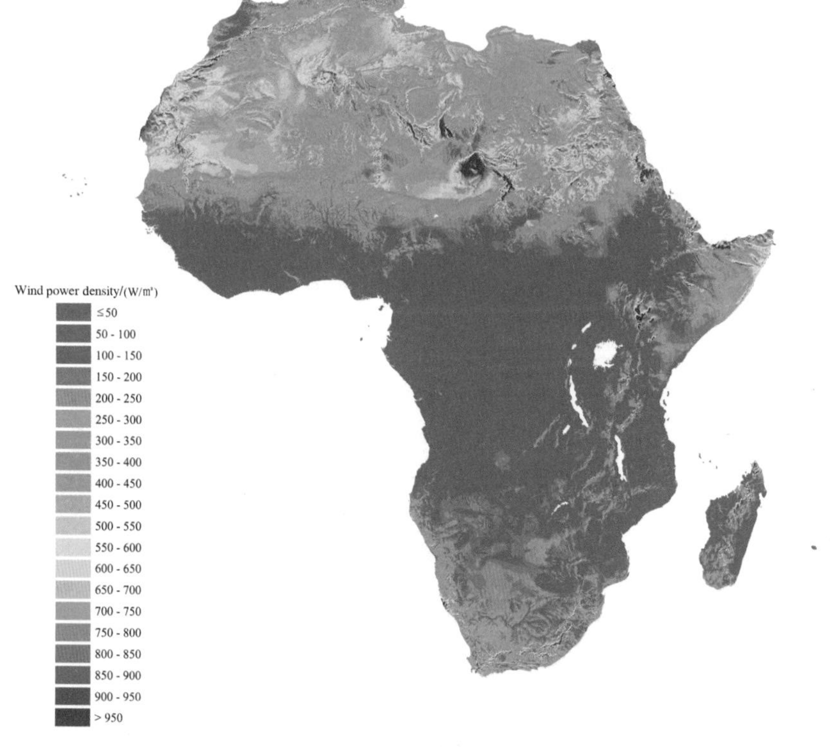


Figure 2. 3 Africa wind power density distribution map at 100 m height
(Data source: Global Wind Atlas)

### 2. 1. 3 Solar Energy Resources

Africa's solar energy resources are widely distributed, among which photovoltaic resources are especially abundant, and the theoretical exploitable quantities of solar thermal and photovoltaic are about 470 PWh/year and 660 PWh/year, respectively, with the exploitable scale accounting for about 40% of the total global solar energy resources. Three-fourths of Africa's land is shining perpendicularly, and most regions have an annual average of more than 2,500 hours of sunshine, especially in most countries in the north, east and south of Africa, where the total annual average horizontal irradiation exceeds 8,000 $MJ/m^2$ (see Figure 2. 4).

### 2. 1. 4 Bioenergy Resources

Africa has abundant bioenergy resources, nearly 90% distributed in the Sahara Desert South region. bioenergy is the most widely used energy in Africa, accounting for 55% of the continent's energy structure, but most of it is used inefficiently for cooking purposes and not effectively converted into electricity. Biogas power generation is one of the main ways of bioenergy power generation, and Africa's hot, dry, and arid climate is a natural advantage for biogas fermentation. It is reported that the biogas production potential in East Africa and North Africa is $1.9 \times 10^6$ $m^3/h$ and $1.3 \times 10^6$ $m^3/h$ respectively, while the potential in West Africa and South Africa is $0.4 \times 10^6$ $m^3/h$ and $0.1 \times 10^6$ $m^3/h$ respectively, and the potential in Central Africa is $0.06 \times 10^6$ $m^3/h$. Currently, there are very few large-scale biogas engineering projects in Africa, with huge potential for

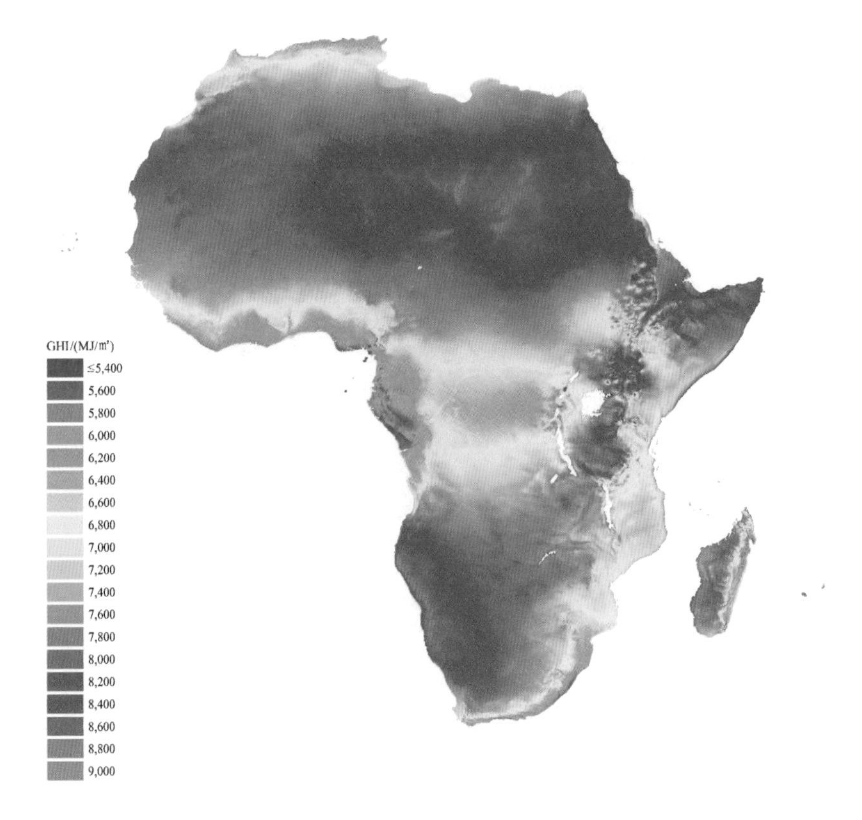


Figure 2. 4 Africa solar resources distribution (Data source: Solargis)

development.

### 2. 1. 5 Geothermal Energy Resources

Geothermal energy resources in Africa are widely distributed in countries such as Djibouti, Ethiopia, Kenya, Rwanda, Tanzania, and Uganda, but the development rate is very low. The potential geothermal energy mainly comes from the East African Rift Valley ( see Figure 2. 5 ), where it has been confirmed that the exploitable volume is about 20 GW. The main body of its geothermal belt is located in the African Plate, running from the Red Sea to the south. After passing through the Ethiopian Plateau, it is divided into east and west branches, with the east branch of the Rift Valley passing through the East African Plateau, and the west

Figure 2. 5 East African Rift Valley

branch of the Rift Valley traveling through Lake Tanganyika. The two Rift Valleys then merge to the south as the Malawi Rift Valley and extend to the south, featuring high heat flow, strong modern volcanism, and extensive fracture activities, with the thermal storage temperatures mostly higher than 200 ℃.

## 2.2 Steady Development of Renewable Energy in Africa

### 2.2.1 The Development of Hydropower

The total installed capacity of hydropower (excluding pumped storage) in Africa grew from 32,991 MW to 37,082 MW during the period 2019 – 2023, an increase of 12.4% in the past five years (see Figure 2.6). The top five countries in Africa in terms of installed hydropower capacity in 2023 are Ethiopia (4,883 MW), Angola (3,729 MW), Congo (DRC) (3,172 MW), Zambia (3,165 MW), and Nigeria (2,851 MW) in sequence (see Figure 2.7).

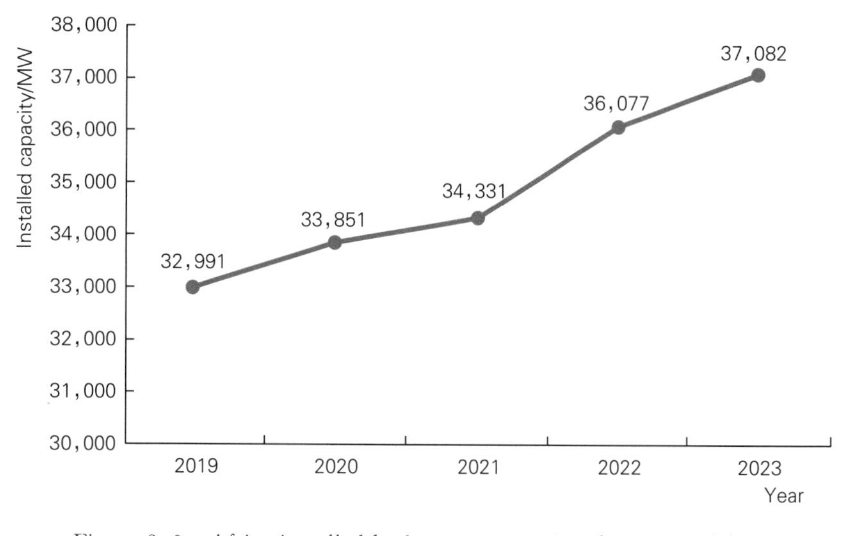


Figure 2.6 Africa installed hydropower capacity change trend from 2019 to 2023 (Data source: IRENA)

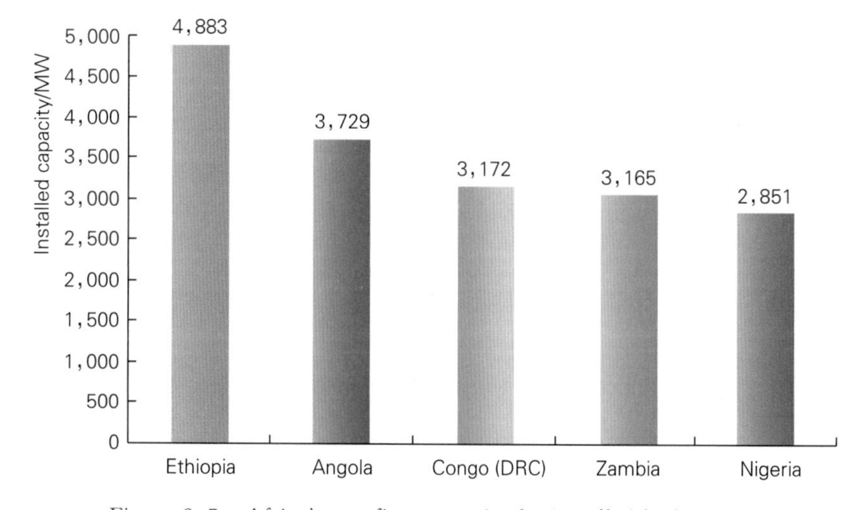


Figure 2.7 Africa's top five countries by installed hydropower capacity in 2023 (Data source: IRENA)

### 2.2.2 The Development of Wind Energy

Africa experiences the total installed wind energyr capacity increased from 5,528 MW to 8,654 MW from 2019 to 2023, a 56.5% increase in the last five years (see Figure 2.8). The top five countries by installed wind energy capacity in Africa in 2023 are South Africa (3,442 MW), Egypt (1,890 MW), Morocco (1,858 MW), Kenya (436 MW), and Ethiopia (324 MW) (see Figure 2.9).

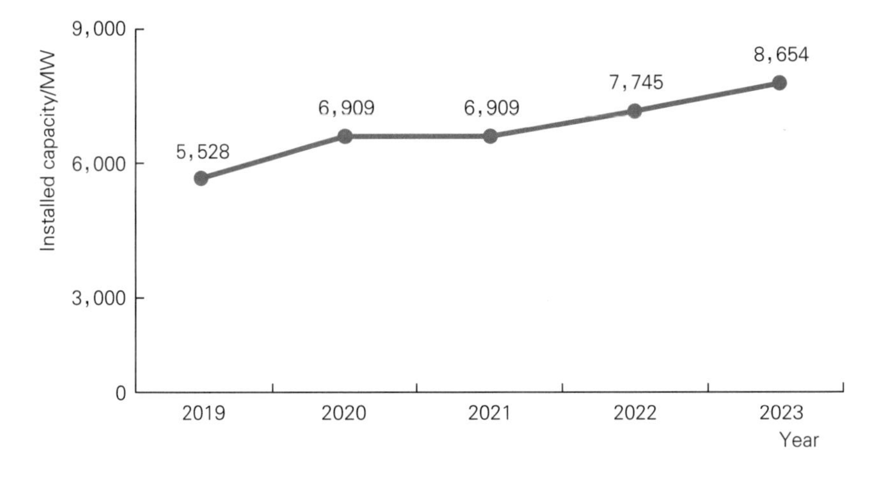


Figure 2.8 Africa installed wind energy capacity change trend from 2019 to 2023 (Data source: IRENA)

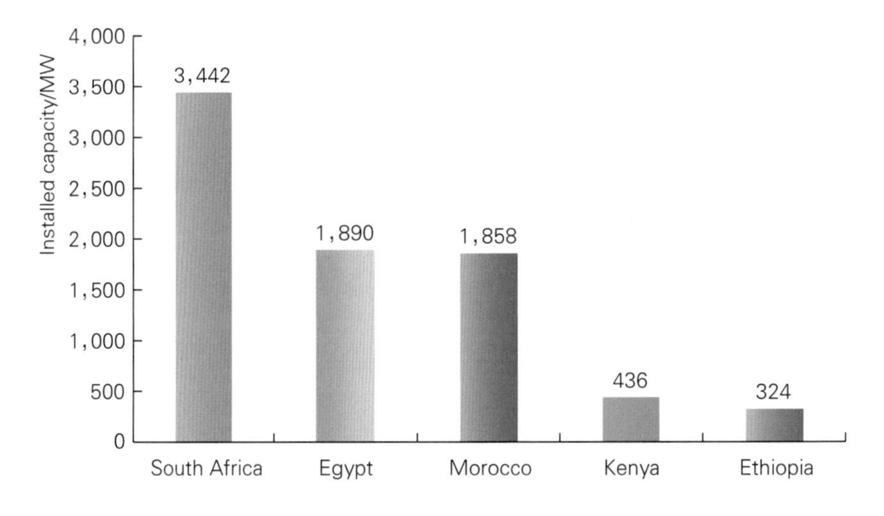


Figure 2.9 Africa's top five countries by installed wind energy capacity in 2023 (Data source: IRENA)

### 2.2.3 The Development of Solar Energy

During the period 2019 – 2023, Africa expanded the total installed solar energy capacity from 9,412 MW to 13,479 MW (including 12,394 MW of PV and 1,085 MW of solar thermal), an increase of 43.2% in the last five years (see Figure 2.10). The top five countries in Africa by installed solar energy capacity in 2023 are

South Africa ( 6, 164 MW ), Egypt ( 1,856 MW ), Morocco ( 934 MW ), Tunisia ( 506 MW ), and Algeria ( 451 MW ) ( see Figure 2. 11 ) .

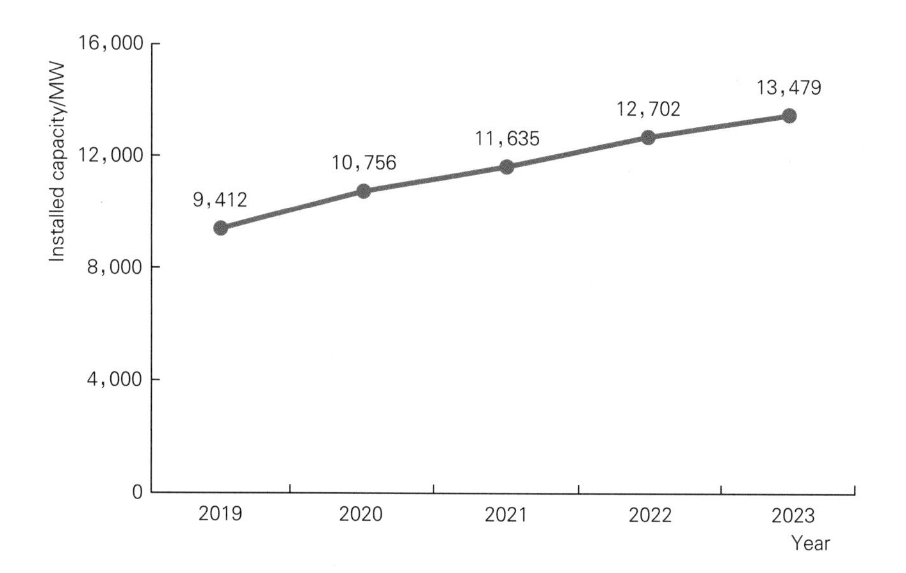


Figure 2. 10 Africa installed solar energy capacity change trend from 2019 to 2023 (Data source: IRENA)

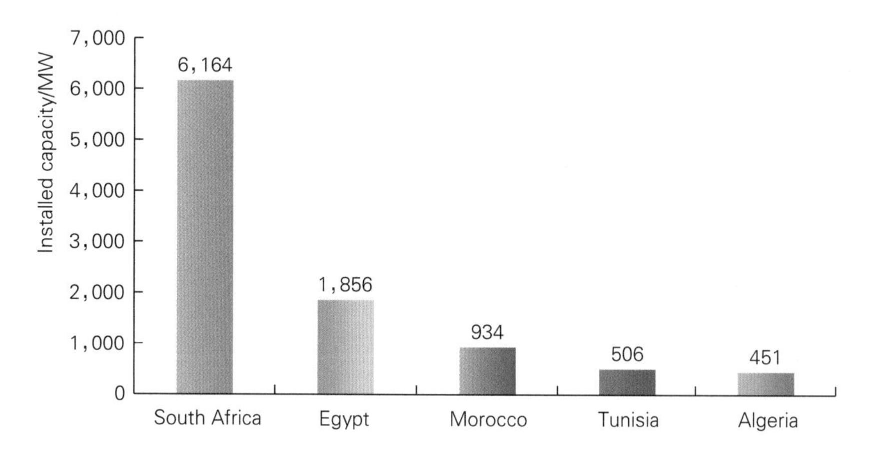


Figure 2. 11 Africa's top five countries by installed solar energy capacity in 2023 (Data source: IRENA)

### 2. 2. 4 The Development of Bioenergy

Africa increased installed bioenergy capacity from 1,807 MW to 1,901 MW during the period 2019 – 2023, an expansion of 5. 2% over the last five years ( see Figure 2. 12 ) . The top five countries by installed bioenergy capacity in Africa in 2023 are Ethiopia ( 310 MW ), South Africa ( 265 MW ), Sudan ( 199 MW ), Egypt ( 131 MW ), and Swaziland ( 106 MW ) ( see Figure 2. 13 ) .

### 2. 2. 5 The Development of Geothermal Energy

Africa grew the total installed capacity of geothermal energy from 691 MW to 991 MW from 2019 to 2023, an

increase of 43.4% in the last five years ( see Figure 2.14 ). Among them, the installed capacity of geothermal energy in Kenya is 984 MW, and it is 7 MW in Ethiopia.

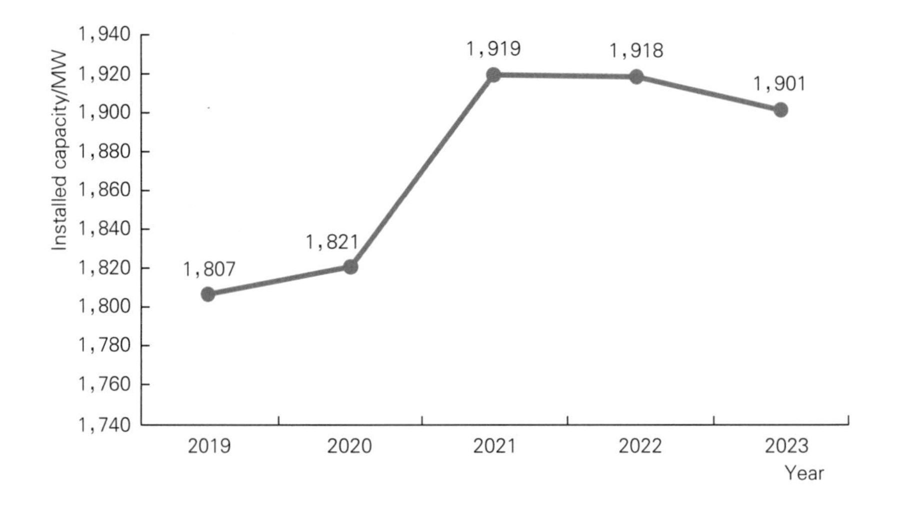


Figure 2.12 Africa installed bioenergy capacity change trend from 2019 to 2023 (Data source: IRENA)

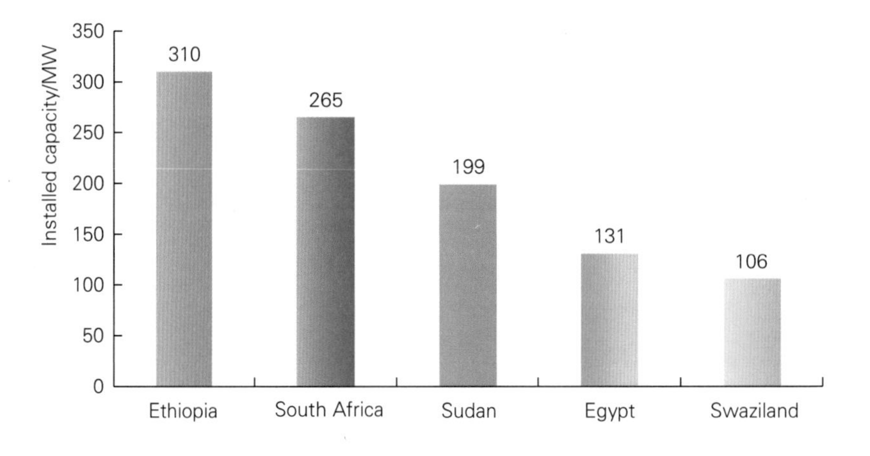


Figure 2.13 Africa's top five countries by installed bioenergy capacity in 2023

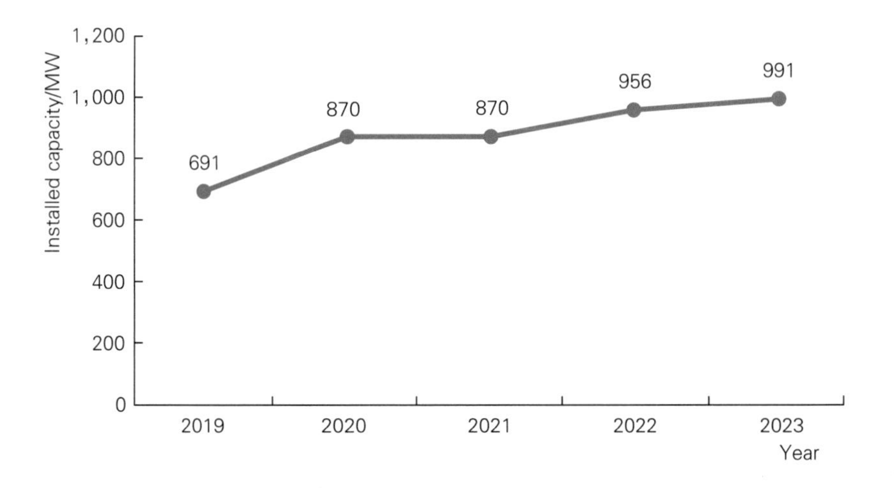


Figure 2.14 Africa installed geothermal power generation capacity change trend from 2019 to 2023 (Data source: IRENA)

### 2.2.6 The Development of Hydrogen

Africa has abundant renewable energy resources, and the advantage of producing hydrogen from renewable energy is obvious. It is reported that as of 2023, at least 12 countries in Africa are carrying out hydrogen energy projects, and 19 countries have developed specific regulatory frameworks or national strategies for the hydrogen energy industry. The African Green Hydrogen Alliance's (AGHA) study in 2022 showed that northern and southern African countries are more suitable for renewable hydrogen production, with about 120,000 – 300,000 tons of hydrogen equivalent export opportunities in Northern Africa, mainly to Europe; and about 100,000 – 220,000 tons of hydrogen equivalent export opportunities in Southern Africa (mainly Namibia and South Africa). The cost of hydrogen production in Africa is expected to fall to less than $1.3 per kilogram in 2050 as capacity costs fall and the electrolyzer supply chain in Africa expands.

(1) North Africa, Sub-Saharan Africa, South America and the Middle East and so on have the greatest potential for green hydrogen globally, with these four regions expected to account for 45% of the global total hydrogen production by 2050, according to a Deloitte study.

(2) Masdar believes that Africa's abundant solar energy and wind energy resources will produce 30 million to 60 million tons of green hydrogen per year by 2050, bringing 1.9 million to 3.7 million jobs in Africa.

(3) The European Investment Bank (EIB) research indicates that Africa has the capacity to achieve an annual green hydrogen production value of 1 trillion euros, and predicts that by 2035, the African continent will be able to produce 50 million tons of green hydrogen annually, with a production cost of less than 2 euros per kilogram.

### 2.2.7 The Development of Energy Storage

Africa has witnessed the installed capacity of wind and solar power an increase of 56.5% and 43.2% respectively in the past five years. Due to the random, volatile and intermittent nature of wind and solar power, the demand for energy storage will become higher in the future. Because many African countries have unstable power systems and serious power shortages and rationing, developing energy storage systems has also served as an effective approach to guarantee the supply of domestic and industrial power. According to predictions from the Global Energy Interconnection Development and Cooperation Organization, the large-scale development and utilization of renewable energy by 2050 will bring about a storage demand of approximately 210 GW and 1,230 GWh for Africa.

Africa is proactively advancing the development of energy storage. Africa had an installed pumped storage capacity of 3,196 MW in 2023, accounting for about 2.3% of the total global installed pumped storage capacity, an increase of 71.5% from 2014. South Africa and Morocco are developing pumped storage, with an installed capacity of 2,732 MW and 464 MW, respectively. In addition, according to the relevant forecasts, Africa's battery energy storage systems market is expected to grow at a CAGR of more than 5.2%

during the period 2024 – 2029. South Africa has undergone a particularly rapid growth in battery energy storage. In November 2023, South Africa announced the inauguration of the largest Battery Energy Storage System ( BESS ) project on the African continent, and according to World Bank forecasts, South Africa's battery storage market is expected to grow to 9,700 MWh by 2030.

### 2.2.8 Summary

In 2023, the total installed capacity in Africa reached 252.8 GW, and fossil energy is still the current main power source in Africa, accounting for about 3/4 of the total installed capacity in Africa. Among renewable energy sources, the installed capacity of hydropower ( excluding pumped storage ) is 37.1 GW, accounting for about 3% of the global installed capacity of hydropower; the installed capacity of wind power and solar power is 8.7 GW and 13.5 GW, respectively, which both accounts for less than 1% globally. Therefore, Africa shows massive potential for renewable energy development.

Africa is accelerating the energy transition and facilitating the development of renewable energy. As shown in Figure 2.15, since 2012, the growth rate of installed capacity of renewable energy ( excluding pumped storage ) in Africa has been significantly greater than the number of installed capacity of fossil energy. In the past five years ( 2019 – 2023 ), the growth rate of total installed capacity of renewable energy ( excluding pumped storage ) in Africa has reached 23.2%, which is 16.8 percentage points higher than the growth rate of installed fossil energy capacity ( 6.4% ) in the past five years. Overall, Africa's abundant renewable energy resources, high development potential and strong willingness for energy transition have paved a good way for the development of renewable energy.

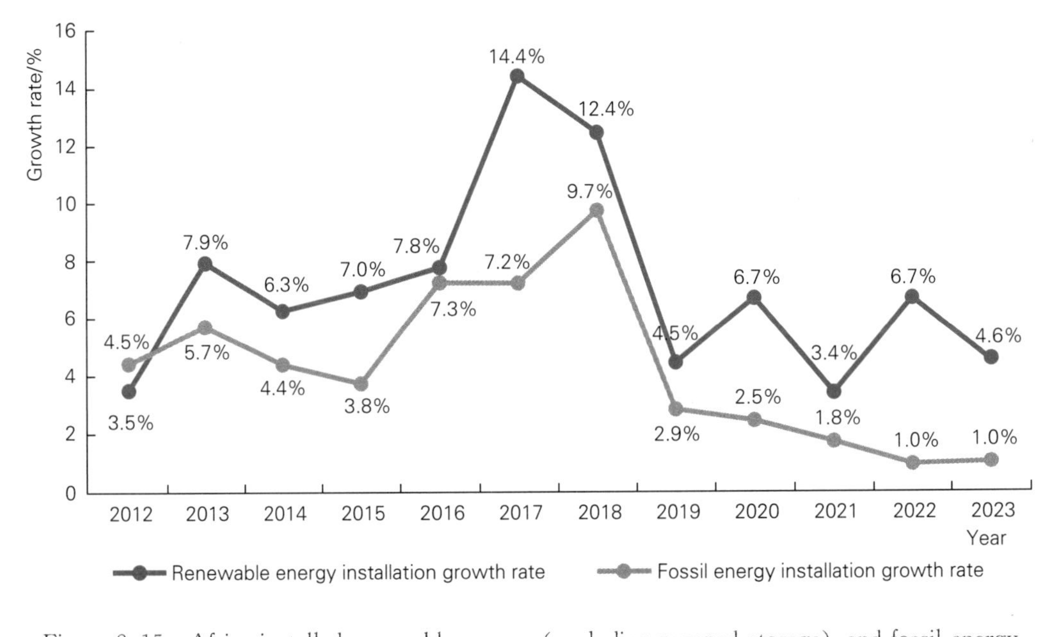


Figure 2.15 Africa installed renewable energy (excluding pumped storage) and fossil energy capacity growth rate comparison from 2012 to 2023 (Data source: IRENA)

## 2.3 Africa Actively Promotes Green Energy Development

### PROGRAMME FOR INFRASTRUCTURE DEVELOPMENT IN AFRICA(PIDA)

Infrastructure development is the cornerstone of Africa's development. In January 2012, PIDA, jointly initiated by the AUC, the AUDA and the AfDB, was adopted at the 18th Ordinary Session of the Assembly of the AU. PIDA aims to fully mobilize relevant resources, promote cross-border infrastructure connectivity and comprehensively modernize infrastructure in Africa.

PIDA is scheduled to be completed by 2040, with a total construction investment of $360 billion, covering four key areas including energy, which focuses on the development of efficient, reliable, affordable and environmentally friendly energy systems and increases access to modern energy services. The program gives priority to accelerating the production and transmission of renewable energy and utilizes the rapid development of renewable energy to provide new pathways for infrastructure development.

### AGENDA 2063

In January 2015, Agenda 2063, initiated by AU leaders, was adopted at the 24th AU Summit. Aiming to formulate a guideline of action for Africa's development planning for the 50 years ahead, Agenda 2063 embodies the beautiful promise of African countries and people focusing on development, looking forward to prosperity and pursuing happiness, painting a grand blueprint for a vibrant and dynamic Africa. It plans to achieve its goals of inclusive and sustainable development in the 50 years during the period 2013 – 2063 and build a regionally integrated, peaceful and prosperous New Africa.

Renewable energy is placed a priority in Agenda 2063, and by 2063, Africa will be globally recognized as an environmentally respectful, eco-conscious continent based on sustainable development and renewable energy. Africa will unlock its full potential for energy production and accelerate its transition from traditional to modern and renewable energy sources, with an aim to ensure that the electricity needs of the majority of its citizens are met, promote economic growth and the eradication of energy-induced poverty and increase the share of renewable energy.

### AFRICA RENEWABLE ENERGY INITIATIVE(AREI)

In December 2015, the AU created AREI under the framework of COP21. It is committed to implementing the requirements of the Paris Agreement, increasing the electrification rate in Africa, accelerating the development of modernized renewable energy sources and improving the accessibility and universality of energy sources. By doing so, it is aimed to make renewable energy available to a larger number of African people, improve the well-being of the African people, drive the African renewable energy revolution and reduce greenhouse gas emissions. All these efforts will lead African countries to the development path in a sustainable and climate-friendly fashion.

AREI facilitates the leapfrogging of renewable energy systems in African countries in support of low-carbon development strategies while enhancing economic and energy security. It helps to achieve sustainable economic development by ensuring access to sufficient renewable, appropriate and affordable energy for all. It would add 300 GW of new renewable and renewable power generation capacity on the continent by 2030, with a focus on hydropower and solar power, complemented by geothermal, wind and biomass power.

## THE AFRICAN SINGLE ELECTRICITY MARKET(AfSEM)

The AU presented a vision of building AfSEM, with aims to integrate power markets across African countries, and promote power trade and interconnection initiatives, increase energy security in the whole of Africa, facilitate the exploits and utilization of the renewable energy and support its economic growth by connecting the electricity infrastructure. Based on the AU's Agenda 2063, AfSEM constitutes a key component of Africa's world-class integrated infrastructure. The establishment of AfSEM will fully accommodate the accessibility of electricity in Africa and further promote sustainable development in the area.

AfSEM aims to provide AU member States with a higher level of energy security, sustainability and competitiveness. It will be the largest single electricity market across the globe, covering 55 member states and benefiting a population of nearly 1.4 billion. AfSEM will effectively respond to Africa's growing electricity demand in the most cost-effective fashion. It is not only a key tool for tapping into the continent's renewable energy potential, but also a powerful gas pedal to promote the realization of full power coverage in Africa, facilitating the rational utilization of power resources in all regions of Africa while contributing to Africa's green development.

## THE AFRICAN CONTINENTAL POWER SYSTEMS MASTERPLAN(CMP)

In order to accelerate the advancement of AfSEM, the AU commissioned AUDA in 2020 to take the lead in initiating the preparation of the CMP, which was formally approved on September 15, 2023 after three years of in-depth research and meticulous planning. The CMP provides a strong technical underpinning for AfSEM, an answer to the full realization of power accessibility and power interconnection in Africa.

The CMP provides an in-depth analysis of the statue quo of Africa's power system, future needs, resource assessment, proper pathways for program implementation, continent-wide contributions, investment requirements, and infrastructure required to enable the energy transportation and trade. It is projected that if Africa's electricity accessibility goal is to be fully met by 2035, Africa's installed renewable energy capacity will hit 750 GW by 2040, and the required electricity investment will be up to about \$ 1.3 trillion during the period 2023 – 2040.

## THE AFRICAN LEADERS NAIROBI DECLARATION ON CLIMATE CHANGE AND CALL TO ACTION

The African Leaders Nairobi Declaration on Climate Change and Call to Action ( hereinafter referred to as the Nairobi Declaration ) was launched on September 6, 2023 at the first Africa Climate Summit. The Nairobi Declaration, which builds on the AU Strategy and Action Plan on Climate Change and Resilient Development

(2022 – 2032), expresses Africa's collective desire to address climate change and underpins Africa's participation in COP28.

The Nairobi Declaration stresses that Africa has a huge potential for renewable energy development as well as abundant natural resources, and reaffirms Africa's willingness to create a favorable environment, put in place policies, promote the necessary investments to unlock the potential of renewable energies, and honor on Africa's climate commitments so as to contribute beneficially to the decarbonization of the global economy. In the Declaration, African leaders recommended that the international community assist Africa in boosting its renewable energy generation capacity from 56 GW in 2022 to 300 GW in 2030, a higher target than the UAE Consensus calls for a three-fold increase by 2030 to 2022.

### DESERT TO POWER INITIATIVE

The initiative, led by the AfDB, is tasked with capitalizing on the Sahel region's abundant solar energy resources and transforming them into renewable energy power plants. The initiative gets involved in 11 countries, including Burkina Faso, Chad, Djibouti, Eritrea, Ethiopia, Mali, Mauritania, the Niger, Nigeria, Senegal and the Sudan, which aims to achieve 10 GW of installed solar power capacity by 2030 through public, private, and grid-connected and off-grid projects, thereby providing access to power for 250 million people.

In addition to solar power, the initiative includes regional power trading, such as the Mauritania-Mali Power Interconnection Project, which aims to enhance regional energy interconnectivity and access to electricity. Furthermore, the initiative is supported by the World Bank Sustainable Energy Fund for Africa to facilitate private sector participation. Not only will it help to reduce dependence on fossil fuels and lower greenhouse gas emissions, but it also represents a vital step in Africa's transition to sustainable energy, which is of great significance for bringing down deforestation and protecting the ecosystem.

### THE AFRICA RENEWABLE ENERGY CORRIDOR(ACEC)

As a regional initiative, ACEC is aimed to accelerate the development of renewable energy potential and facilitate cross-boundary trade of renewable energy within EAPP and SAPP. The initiative builds on the political commitment of African leaders to strengthen regional institutions and transmission infrastructure, with a view to create large competitive markets and reduce costs across production sectors.

By creating a broader regional electricity market, ACEC will attract investment to meet 40% to 50% of the electricity needs in EAPP and SAPP regions by 2030. ACEC needs to invest up to $25 billion in a single year in power generation by 2030 and an additional $15 billion per year in grid infrastructure. We are working together to diversify resources, improve energy security and create investment opportunities and job growth. Expanding renewable energy application also brings a comprehensive opportunity for Africa to keep carbon-intensive infrastructure lock-in at bay and move rapidly towards a low-carbon future.

### AFRICAN MINI-GRID PROGRAM(AMP)

AMP (2022 – 2027) is a government-led technical assistance project totaling $50 million, which is funded by

the Global Environment Facility ( GEF ) and implemented by the United Nations Development Programme in partnership with the Rocky Mountain Institute ( RMI ) and AfDB. The project takes the mission of stimulating the solar battery mini-grid market to increase access to electricity in 21 countries. The goal is to bring the developmental benefits of energy access to a wide range of communities across the continent by focusing on supporting the productive purpose of energy. Thereby, it makes contribution to socio-economic development through improving quality in sectors that require energy investment, such as agriculture, health care, education, and small businesses.

The 21 countries include Angola, Benin, Burkina Faso, Burundi, Chad, Comoros, the Democratic Republic of the Congo, Djibouti, Ethiopia, Swatini, Liberia, Madagascar, Malawi, Mali, Mauritania, Niger, Nigeria, Sao Tome and Principe, Somalia, the Sudan and Zambia.

### SCALING SOLAR PROGRAM

The Scaling Solar Program, initiated by the World Bank through IFC, assists African countries in installing PV power plants in the form of tenders, with the goal of encouraging private companies to invest in PV to supply electricity to the relevant national grids. Scaling Solar institutes a standardized methodology for project development, including the bidding process, financing, and risk management. Solar projects have been successfully implemented in countries such as Zambia, Senegal and Ethiopia, significantly reducing the cost of solar power generation in these regions.

# 3

# China-Africa Renewable Energy Cooperation History and Achievements

> "China and African countries have established multi-disciplinary cooperation, which lays a good foundation for helping African countries boost innovative green development."
>
> —UN Secretary-General António Guterres

Since the first generation of leaders of the PRC and the old generation of African politicians laid down the foundation of friendship, the two sides have followed of a road of win-win cooperation with mutual respect, love and support. Over the past 74 years, China and Africa have been in the same boat and moved forward hand in hand. China is committed to consolidating mutual trust between China and Africa in political terms, deepening pragmatic cooperation in multiple fields and providing all the help it can for Africa's peace and development. As a result, the cooperation between China and Africa has always been at the forefront of international cooperation with the continent. Under the diplomatic leadership of the heads of state and government of Africa and China and the guidance of top-level design, the two sides have established increasingly perfect cooperation mechanisms in renewable energy, technological innovation has kept abreast of the times, bringing project cooperation to fruition, and the mode of cooperation has evolved from early main foreign aid and engineering contracting to integration of investment, construction and operation. In 1964, the Jinkang Hydropower Station in Guinea, which was built in support of the Chinese government, was the first renewable energy cooperation project between the two sides. Over the past decades, Chinese enterprises have actively responded to the strategy of "going global" and built a host of renewable energy projects in Africa, These projects include some key landmark projects such as the Inauguration Du Barrage Hydroelectrique De Soubre in Côte d'Ivoire, the De Aar Wind Power Station in South Africa, and the Sakai Photovoltaic Power Station in the Central African Republic. The results of cooperation in renewable energy have been spread all over Africa, improving the conditions for economic and social development in Africa and bringing tangible benefits to the people of both sides.

## 3.1 China-Africa Cooperation Has a Last-standing History

After the founding of the PRC, China and Africa firmly supported each other in the anti-imperialist and anti-colonial struggles for national liberation, and forged an unbreakable brotherhood friendship. At the Bandung Conference in 1955, Premier Zhou Enlai put forward the policy of "Seeking Common Ground while Reserving Differences",

laying the foundation of solidarity and cooperation between China and Asian and African countries. Subsequently, China and Egypt established diplomatic relations in May 1956, ushering in a new era of Sino-African official exchanges. On this basis, by the end of 1963, China had established diplomatic relations with 12 African countries, and from 1963 to 1964, Premier Zhou Enlai and other Chinese leaders visited 10 African countries, which was hailed as a "pioneering journey" in the development of China-Africa relations and further deepened the ties between the two sides. These exchanges not only marked the transformation of China-Africa relations from civil to official, but also laid the foundation for subsequent China-Africa cooperation on an equal and sincere basis. The Five Principles of Peaceful Coexistence advocated by Premier Zhou Enlai became the core principle guiding the relations between the two sides, which is of far-reaching significance globally.

In the 1970s, China forged a more friendly and cooperative relations with Africa. China established diplomatic relations with a total of 25 African countries and provided considerable non-reimbursable economic assistance for African countries, the total amount of which reached 200 million US dollars from 1965 to 1969, especially supporting the development of Tanzania, Zambia, and Guinea, etc. The construction of Tanzanian Railway assisted by China was completed in 1976, which became an important transportation artery connecting East Africa and Central and South Africa. In 1976, the Tanzanian-Zanzibar Railway was completed, becoming an important transportation artery connecting East and Central Africa, which is one of China's largest foreign-aid projects with a total length of 1,860. 5 kilometers. During this period, despite the domestic difficulties and challenges that China faced I, China-Africa relations maintained a good momentum of development. China provided relentless support for national independence and liberation movements in Africa, while African countries firmly advocated China' stance in the international arena, especially in supports of the restoration of China's legitimate seat in the United Nations. This history bears witness to the profound friendship and mutually supportive partnership between China and Africa.

Since the introduction of the reform and opening up, China-Africa relations entered a new stage characterized by pragmatic cooperation. In the 1980s, the turnover of China's labor cooperation and contracted projects in Africa exceeded US $ 2. 5 billion, while in the 1990s the average annual trade volume between China and Africa grew from more than US $ 1 billion to US $ 6. 48 billion in 1999. During this period, Chinese leaders visited a host of African countries and proclaimed four principles for economic and technical cooperation with African countries, namely, equality and mutual benefit, pragmatism, multiple orms and common development, which marked a new level taken to in China-Africa economic and technical cooperation. In the meantime, China-Africa relations have shifted from gratuitous aid to cooperation in trade and infrastructure construction, and Africa has acted as a crucial partner in the implementation of China's multilateral diplomacy. The two sides shared robust economic complementarities and their economic and trade cooperation achieved swift progress.

In 2000, the FOCAC came into being, leading China-Africa friendship and pragmatic cooperation to achieve leapfrog development, and China-Africa relations entered a new phase of institutionalized cooperation. In October 2000, the first FOCAC was held in Beijing , with the aim of constructing a new international political

and economic order and promoting China-Africa cooperation, and it adopted *the Beijing Declaration and the China-Africa Economic Cooperation and Social Development Programme*. In 2006, Hu Jintao visited three African countries and delivered a speech in the Nigerian Parliament, emphasizing the friendly relations between China and Africa and putting forward a five-point proposal for the development of a new type of cooperative partnership between China and Africa. In November of the same year, the Beijing Summit was successfully convened, which marked one of the largest foreign affairs events since the founding of the PRC, and important documents guiding the development of China-Africa relations were adopted.

In 2013, President Xi Jinping made his first visit to Africa after assuming the presidency, emphasizing that China and Africa have always been a community of a shared future, and putting forward the principle of upholding justice while pursuing shared interests and adhering to the principle of sincerity, real results, affinity, and good faith, which charted the way and provided a fundamental guideline for cooperation with Africa in the new era. At the Johannesburg Summit of the FOCAC in December 2015, President Xi Jinping proposed elevating China-Africa relations to a comprehensive strategic partnership. In September 2018, President Xi Jinping proposed building a closer China-Africa community of a shared future at the Beijing Summit of the FOCAC. In November 2021, President Xi Jinping proposed to work together to build a community of a shared future for China-Africa in a new era at the 8th Ministerial Conference of FOCAC. Over the past decade, President Xi Jinping has paid five visits to Africa, attended two summits of the FOCAC, promoted the deep implementation of China-Africa cooperation under the framework of the Belt and Road Initiative and the FOCAC, presided over the China-Africa Unity Against Epidemic (CAUAU) Special Summit and the China-Africa Leaders' Dialogues. With these efforts, the President summarized the spirit of China-Africa friendship and cooperation, and personally charted the development of the relationship between China and Africa in the new era. He has also summarized and refined the spirit of China-Africa friendship and cooperation, personally charted the course for the development of China-Africa relations in the new era, which has continued to benefit people on both sides and written a new chapter of China-Africa friendship and cooperation.

The FOCAC has gone through more than 20 years of history and has become a pacesetter leading China-Africa cooperation and even international cooperation with Africa. China-Africa relations have achieved a "three-stage leap" from a new partnership to a new strategic partnership to a comprehensive strategic partnership. Ranging from the Ten Major Cooperation Projects to the Eight Major Initiatives to the Nine Projects, China-Africa cooperation has been constantly expanding and upgrading, yielding fruitful results. Trade volume has increased from $10.5 billion in 2000 to $282.1 billion in 2023, a nearly 26-fold increase. China's investment in Africa has grown from less than $500 million in 2000 to more than $40 billion at present. Over the years, China-Africa cooperation has built and upgraded nearly 100,000 kilometers of roads, more than 10,000 kilometers of railroads, nearly 1,000 bridges and nearly 100 ports. China-Africa exchanges in multiply fields have been more active than ever before, with exchanges and mutual learning in tourism, culture, youth and media blossoming, and the foundation of friendship between the two sides has been more consolidated.

## 3.2 Fruitful Cooperation on Renewable Energy in China and Africa

### 3.2.1 Continuous Improvement in Mechanism Development

Mechanisms in renewable energy between China and Africa are becoming increasingly perfect. Under the guidance of the Belt and Road Initiative and the Forum on China-Africa Cooperation, the development strategies of the two sides in the energy sector have been deeply dovetailed; in the context of South-South cooperation in addressing climate change, China actively participates in global climate governance; under the United Nations, the G20, the APEC and other multilateral mechanisms, China and the international community have worked together to consolidate and deepen regional energy; under the framework of the China-AU Energy Partnership, China has actively promoted practical cooperation in the areas of policy and information exchange, capacity building and tripartite cooperation. By the end of 2021, China had established bilateral committees, diplomatic consultations or strategic dialogues mechanisms with 21 African countries and the AU Commission, as well as economic and trade joint (hybrid) committees with 51 African countries. Under the aforementioned bilateral cooperation mechanisms, China and African countries have been promoting the incubation and implementation of renewable energy cooperation projects.

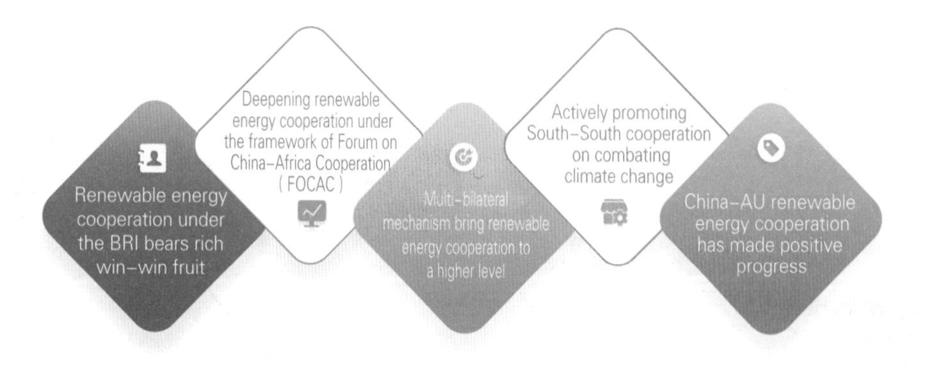


#### Renewable Energy Cooperation under the BRI Bears Rich Win-win Fruits

In September and October 2013, President Xi Jinping proposed the launch of the "New Silk Road Economic Belt" and the "$21^{st}$-Century Maritime Silk Road" ("One Belt and One Road") Cooperation Initiatives. In the past ten years (2013 - 2023), China has signed cooperation documents on the building of the BRI with more than 150 countries and more than 30 international organizations. As of the end of 2023, 52 African countries as well as the AUC have signed cooperation documents on the BRI with China, marking Africa have been one of the most important continents participating in the BRI. Under the guidance of the Initiative, the two sides have constructed a number of BRI landmark green projects, which have vigorously facilitated the development of the energy industry in the African countries concerned and provided reference and demonstration for other developing countries. With the in-depth advancement of the BRI, Africa and China have continued to

consolidate policy communication in renewable energy and further strengthen the consensus on cooperation. In the meantime, Africa has been speeding up the integration of renewable energy resources and unleashing the energy market, leading to a growing scale of development.

During the Second BRI Summit Forum, the BRI Energy Partnership was established in Beijing On April 25, 2019. The organization is committed to promoting mutually beneficial energy cooperation, helping countries to jointly tackle the problems over the course of energy development, and seeking common development and shared prosperity. It is an important measure to steer the BRI towards greater success. At present, 33 member states are establishing the "BRI" energy partnership, including 9 African countries. In 2020, China and the AU signed a MoU on the cooperation working coordination mechanism of the jointly building BRI. The AU was the first regional organization to sign a cooperation program under the BRI with China and took the lead in setting up a working coordination mechanism. The two sides are working together on cooperation in the field of renewable energy and other infrastructures under the framework of the BRI, which has effectively advanced the regional economic integration of the African continent.

Infrastructure construction is a strategic priority area in the BRI. In June 2023, the 3rd China-Africa Infrastructure Cooperation Forum was convened, in which 19 contracts were signed in the field of engineering contracting and engineering investment to Africa. It is involving infrastructure cooperation in energy, communication, industry and agriculture and other specialized fields in many African countries such as Nigeria, Kenya, Ghana, Uganda, Côte d'Ivoire, Republic of the Congo, Egypt, Morocco, Zimbabwe, Niger, South Africa. The total amount of the signed projects exceeded 2. 9 billion dollars, with a total of 32 contracting units.

The South African ambassador to China: The achievements of the BRI cooperation meet the development needs of Africa

China-Africa Infrastructure Cooperation Forum Projects Signing Ceremony

### Deepening Renewable Energy Cooperation under the Framework of Forum on China-Africa Cooperation (FOCAC)

Founded in October 2000, the FOCAC is a collective dialog mechanism between China and African countries on the basis of equality and mutual benefit, which is committed to further solidifying China-Africa friendly

cooperation in the new situation and seeking common development. Since the inception of FOCAC, the two sides have focused on the development of wind power, photovoltaic and other renewable energy industries and the construction of power infrastructure, which effectively advances Africa's green transformation and sustainable development, achieving a series of cooperative results. Under the mechanism of the Forum, China has implemented hundreds of renewable energy and green power generation projects in Africa, so as to promote the complementarity of the two sides' advantageous resources and facilitate mutual benefits and win-win cooperation.

At the Johannesburg Summit of the FOCAC in 2015, China put forward the "10 major cooperation projects" in China and Africa, promising that "China will support Africa to bolster its capacity in green, low-carbon and sustainable development and assist the implementation of 100 clean energy projects in Africa". The 2018 Beijing Summit adopted *the Beijing Declaration* and *Beijing Action Plan (2019 -2021)*. These two fruitful documentations propose that Africa and China strengthen policy dialogues and technical exchanges in the fields of energy and resources, align the development strategies of energy and resources, carry out joint research, and jointly develop locally adapted and highly operational energy development plans. The 8th Ministerial Conference of the FOCAC in 2021 adopted the Dakar Declaration and the Dakar Plan of Action (2022 – 2024). According to these two fruitful documentations, China and African countries will strengthen practical cooperation in the energy sector under the framework of the China-AU Energy Partnership, jointly improve the level of electrification in Africa, increase the proportion of renewable energy, and gradually tackle the problem of energy accessibility so as to push the sustainable energy development on both sides. China will strengthen docking with the development strategies of the AU and African countries, and support the African side in accelerating the implementation of the AU's Agenda 2063, so as to realize autonomous and sustainable development on an earlier data.

Beijing Summit of the Forum on China-Africa Cooperation in 2018

The Third China-Africa Economic and Trade Expo in 2023

President Xi Jinping announced the establishment of the China-Africa Economic and Trade Expo at the 2018 Beijing Summit of the FOCAC. In June 2023, the third China-Africa Economic and Trade Expo was convened with the theme of "Seek Common Development, Share the Future" to implement the spirit of the 8th

Ministerial Conference of the FOCAC. At the meeting, 120 projects amounting to 10.3 billion dollars were signed, and 99 integration projects worth 8.7 billion dollars were published, of which 74 integrated projects were from 11 African countries, the largest number of events ever. Energy cooperation projects are one of the key signing objects, including Zimbabwe's 100 MW photovoltaic power station project, Niger's 250 MW wind, light and storage smart energy park project, Egypt's Port Said green ammonia project, and South Africa's 500 MW photovoltaic and 1,000 MWh energy storage project, among the others.

### Multi-bilateral Mechanisms Bring Renewable Energy Cooperation to a Higher Level

Under the UN framework and important multilateral mechanisms such as G20, APEC, IRENA and BRICS, Africa and China work together to push green and sustainable development. Africa and China have continuously diversified and improved intergovernmental dialogues, consultations and cooperation mechanisms, brought full play to the role of coordinated arrangement, and pushed the all-round development of China-Africa cooperation in multiple fields. Under the bilateral cooperation mechanisms in the energy sector, such as the China-Morocco Executive Committee on Energy Cooperation and the Energy Committee of the China-South Africa Economic and Trade Association, China has brought its cooperation in renewable energy with key regions and countries to a higher level.

China has long attached great importance to enhancing electricity access in Africa. As early as 2015, China, as one of the founding members of the G20, was in firm support of the G20 Energy Access Action Plan: Voluntary Collaboration on Energy Access, which takes upgrading electricity access in sub-Saharan Africa as an important element. In November 2022, at the seventeenth G20 Summit, China took the lead in publicly supporting the AU's accession to the G20 initiative, pushing the international community to further focus on the issue of electricity access in Africa and promote renewable energy cooperation between Africa and China under multilateral frameworks such as the G20. At the 15th BRICS Summit, South Africa and China signed a Letter of Intent on Deepening BRI Cooperation. The energy enterprises from both sides signed a memorandum of related strategic cooperation and have actively participated in South Africa's renewable energy project investment. On June 7, 2021, NEA and IRENA signed a MoU, in which the two sides agreed to carry out cooperation in three areas, namely, energy transition strategies and policies, promotion and application of renewable energy technologies, and assistance and support for the development of renewable energy in other countries; in the same year, IRENA, in cooperation with relevant organizations, supported the AUDA to develop CMPs, review and reconsider power generation options to maximize socio-economic benefits while minimizing emissions.

In April 2023, NEA was invited to send missions to Namibia and Angola, and the parties concerned made exchanges of views on consolidating cooperation in the energy sector. During the visit of the President of Algeria to China, the Chinese and Algerian energy authorities signed a MoU on China-Algeria Renewable Energy Cooperation to promote cooperation in capacity building, joint research and demonstration projects, etc, Africa and China have carried out capacity-building training in multiple forms under multi-bilateral mechanisms, cultivating a large number of talents for African countries. It is reported that during the period 2015 – 2018

alone, China has trained 200,000 various types of vocational and technical personnel for Africa, and provided 40,000 quotas for training in China and 2,000 educational quotas for academic degrees and diplomas. According to incomplete statistics, since 2018, China has set up 12 Luban workshops in 11 African countries, including Djibouti, Egypt, South Africa, Kenya, Nigeria and Côte d'Ivoire, to share high-quality education with Africa in multiple fields including new energy and to help the AU accomplish the goal of "empowering 70% of young people a skill" .

Meeting between Energy Authorities of China and South Africa in 2023

Ministerial Seminar on Capacity Building for State Governance and Macroeconomic Planning in Zambia in 2024

### Actively Promoting South-South Cooperation on Combating Climate Change

By the end of 2023, China had signed 50 memorandums of understanding on South-South cooperation on combating climate change with 41 developing countries, which includes 18 South-South cooperation documents on combating climate change with 16 African countries such as Nigeria, Ethiopia and Benin. In doing so, China remains committed to providing support to developing countries including African countries by cooperating in the construction of low-carbon demonstration zones, carrying out climate change mitigation and adaptation projects, and organizing capacity-building training.

Before the 8th Ministerial Conference of the FOCAC in 2021, Africa and China have jointly developed *China-Africa Cooperation Vision 2035*, and as the first three-year plan of the *China-Africa Cooperation Vision 2035*, China will work closely with African countries to jointly implement "nine programs-the green development" The Vision proposes that China-Africa renewable energy cooperation should be transformed into clean and low-carbon, creating a new model of green development, and realizing China-Africa ecological co-construction. At the meeting, the two sides adopted *the Declaration on China-Africa Cooperation on Combating Climate Change*, proposed that both sides shall strengthen China-Africa cooperation to address climate change, and implement practical cooperation projects in renewable energy and other areas so as to jointly tackle the challenges posed by climate change. The document points out that Africa and China are ready to further strengthen South-South cooperation on combating climate change, broaden the areas of cooperation, and

carry out pragmatic cooperation projects in renewable energy and other areas.

To further implement *the Declaration on China-Africa Cooperation on Combating Climate Change*, China announced at the Africa Climate Summit that it would develop and implement the "Africa Light Belt" project, a South-South cooperation project to fight against climate change, focus on Africa's demand for photovoltaic resources and renewable energy, leverage China's strengths in the photovoltaic industry, and adopt the approach of "material assistance, exchange and dialogue, joint research and capacity building" to build a cooperation demonstration belt in the utilization of photovoltaic resources in Africa and China. In addition, China has also provided solutions for Africa's response to climate change through the implementation of climate change mitigation and adaptation projects and the joint construction of low-carbon demonstration zones. During the COP28, China signed the first "African Light Belt" project document with Chad and a MoU with São Tomé and Príncipe on the "African Light Belt" project, the South-South cooperation on combating climate change.

China and Seychelles signed a MoU on the construction and implementation of a low-carbon demonstration zone for South-South cooperation on combating climate change in 2021

China announced at the Africa Climate Summit in 2023 that it would develop and implement the "African Light Belt" project

### China-AU Renewable Energy Cooperation Has Made Positive Progress

From April to May 2018, the NEA commissioned two expert groups to six countries, including Ethiopia, Kenya, South Africa, Egypt, Nigeria and Côte d'Ivoire, to carry out research. This move lays the foundation for promoting China-Africa energy cooperation at multiple levels and in diverse forms and boosts the establishment of an intergovernmental energy cooperation mechanism among China and the AU and African countries. In September 2018, the NEA and AUC signed a MoU on strengthening cooperation in the energy field. The two sides agree to jointly promote comprehensive cooperation between China, African countries and African regional organizations in the renewable energy sector under the guidance of the principle of "extensive consultation, joint contribution and shared benefits".

In October 2021, the NEA and AUC signed and established the China-AU Energy Partnership, in which the two sides agreed to carry out pragmatic cooperation in the areas of policy and information exchanges, capacity

building, project cooperation and tripartite cooperation. The 8th Ministerial Conference of the FOCAC adopted the *Dakar Declaration and Dakar Action Plan* ( *2022 – 2024* ), both acknowledge the China-AU Energy partnership is an important part of pragmatic cooperation between Africa and China in the energy sector. The two sides have already held a series of seminars and conferences, capacity-building training, project promotion and other activities relying on the China-AU Energy Partnership.

In 2018 NEA commissioned a group of experts to exchanges with Ministry of Electricity and Renewable Energy of Eygpt

Inspection of Dadu River Angu Hydropower Station during the first capacity building session under the framework of the China-AU Energy Partnership in 2022

In terms of promoting exchanges and seminars, China and African government departments, financial institutions, and energy enterprises make exchanges of technical experiences and share project opportunities through a series of seminars and conferences, while calling for the international community to focus on the development of renewable energy in Africa. With regard to facilitating capacity building, relevant organizations are pulled together to organize capacity-building training programs on topics related to renewable energy and power grids so as to enhance the management and technical level of African energy industry practitioners. In the aspect of deepening project cooperation, a project cooperation seminar was held under the theme of "Sharing African Energy Development Opportunities and Promoting Pragmatic Cooperation for China-Africa Projects". It is aimed to help Chinese enterprises to seek cooperation opportunities in energy projects in Africa, and actively participate in the investment and construction of PIDA projects and energy projects in African countries. As for the construction of "small and beautiful" demonstration projects, innovative cooperation project cases and technical solutions are promoted through energy cooperation seminars, accelerator programs and COP28-related activities, thereby advancing the implementation of "small and beautiful" projects in Africa.

During COP27, the relevant units of China and Africa launched *Egypt Initiative on Developing Clean Energy to Improve Electricity Access in Africa* on the basis of the China-AU Energy Partnership. Both sides call on the international community to further focus on the problem of people without access to electricity in Africa and to take more powerful and more sustainable measures to promote the development of renewable energy in Africa

based on the principles of diversity, equity and inclusiveness, so as to contribute to enhancing electricity accessibility in Africa.

2023 COP28 China Corner announced the Official launch of the China-Africa Energy Innovation Cooperation Accelerator Project

2023 The First Project Promotion Conference of Energy Cooperation under the framework of China-AU Energy Partnership

### 3.2.2 Strategic Planning Continues to Advance

❖ Vision and Proposed Actions Outlined on Jointly Building Silk Road Economic Belt and 21st-Century Maritime Silk Road

In December 2020, the Director General of the National Development and Reform Commission (NDRC) and the Chairperson of the AU Commission (AUC) signed *the Cooperation Plan on Jointly Promoting the Silk Road Economic Belt and the 21st Century Maritime Silk Road between the Government of the PRC and the AU*, the first agreement of its kind signed between China and a regional organization. It focuses on the areas of policy communication, facility connectivity, trade facilitation, financial integration and people-to-people connectivity, clearly defines the contents of cooperation and key cooperation projects, and puts forward the schedule and road-map.

The signing of *the Cooperation Plan on Jointly Promoting the Silk Road Economic Belt and the 21st Century Maritime Silk Road between the Government of the PRC and the AU* will effectively promote the alignment of the BRI with the AU's Agenda 2063, strengthen the complementarity of the two sides' strengths, jointly address global challenges, propel the high-quality development of the BRI, which creates new opportunities for global cooperation, and injects more impetus to common development. With the signing of the Plan, China and the AU Commission will set up a working coordination mechanism for jointly building the BRI cooperation to further promote the implementation of the Cooperation Plan.

❖ China-Africa Cooperation Vision 2035

The China-Africa Cooperation Vision 2035 was adopted during the 8th Ministerial Conference of the FOCAC in 2021 in order to further strengthen China-Africa strategic consensus, consolidate China-Africa comprehensive

strategic partnership, reinforce comprehensive pragmatic cooperation, and make a good medium-and long-term planning for China-Africa cooperation. Aligning with China Vision 2035, the United Nations 2030 Agenda for Sustainable Development, the AU's Agenda 2063 and the development strategies of African countries, the vision defines the overall framework for China-Africa cooperation in the coming 15 years and envisions the future for China-Africa cooperation in multiple fields in 2035.

The China-Africa Cooperation Vision 2035 indicates that the two sides will work together to build a new pattern of transformation growth, accomplish China-Africa industrial co-promotion, and beef up energy cooperation for a cleaner, low-carbon transition. China encourages Africa to increase the proportion of renewable energy utilization such as hydro energy and nuclear energy and proactively develop renewable energy such as solar energy, wind energy, geothermal energy, biogas, tidal current and wave energy based on actual conditions and energy needs of the individual country, in order to provide a stable and affordable power supply for the remote areas in Africa through distributed power supply technology, and promote the development of the photovoltaic industry.

### Nine Programs—The Green Development Program

President Xi Jinping announced at the 8th Ministerial Conference of the FOCAC that, as the first three-year plan of the China-Africa Cooperation Vision 2035, China will work closely with African countries to jointly implement the "Nine Programs—The Green Development Program". The Green Development Program proposes that "China will assist in implementing ten green environmental protection and climate change response projects for Africa, support the construction of the 'Great Green Wall in Africa', and build low-carbon demonstration zones and climate change adaptation demonstration zones in the region." China is focusing on cooperation with Africa in providing climate-change assistance and strengthening green cooperation, which means assisting African countries in implementing environmental protection and climate change response projects, working with relevant African countries to build low-carbon demonstration zones and climate change adaptation demonstration zones and carrying out South-South cooperation projects on climate change mitigation and adaptation.

### Declaration on China-Africa Cooperation Combating Climate Change

We are undergoing a prominent global challenge posed by climate change at present. In order to jointly fight against climate change, Africa and China have adopted *the Declaration on China-Africa Cooperation Combating Climate Change* during the 8th Ministerial Conference of the FOCAC. *The Declaration on China-Africa Cooperation Combating Climate Change* proposes the establishment of a China-Africa strategic partnership to address climate change for a new era, turning a new page in China-Africa green and low-carbon development, and highlighting the determination of Africa and China to further strengthen exchanges and cooperation on combating climate change under the framework of South-South cooperation and the BRI in the future. *The Declaration on China-Africa Cooperation Combating Climate Change* shows that the two sides unanimously advocate sustainable development that is innovative, coordinated, green, open and shared, and are getting ready to further strengthen South-South cooperation between Africa and China in tackling climate change, expand cooperation areas, and carry out

pragmatic cooperation projects in multiple areas such as renewable energy.

### 3.2.3 Expanding and Deepening Project Cooperation

China-Africa has established an increasingly close relationship in renewable energy cooperation and the two sides have successfully cooperated on a large number of high-quality renewable energy power projects. According to IEA statistics, during the period 2010 - 2015, Chinese-funded enterprises, as one of the most important contractors on the African continent, undertook installed power capacity in various types of projects accounting for about one-third of the new capacity in sub-Saharan Africa; during the period 2010 - 2020, Chinese enterprises constructed or were constructing more than 200 power projects in sub-Saharan Africa, involving 37 countries and regions, with a total of about 17 GW of installed capacity, which is equivalent to 10% of the existing installed capacity in sub-Saharan Africa. The installed capacity of renewable energy projects accounted for more than half of the total.

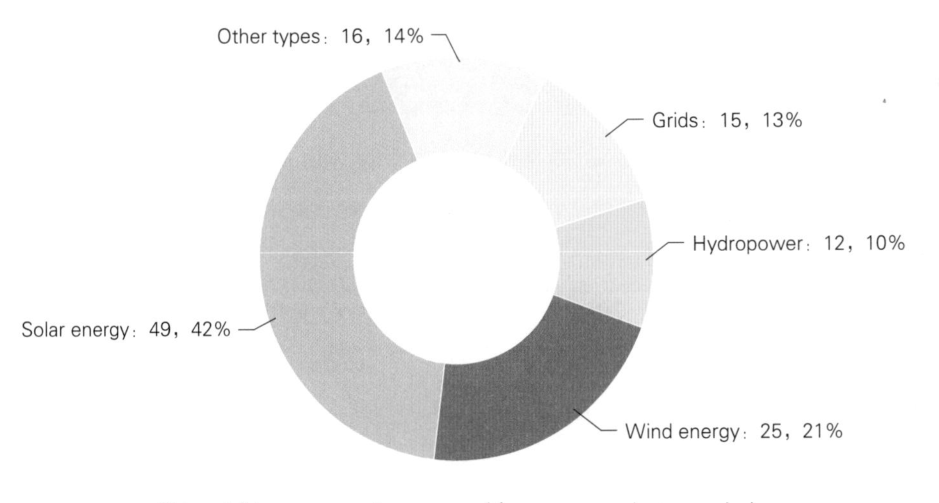


China-Africa cooperation renewable energy projects statistics

In order to further understand the cooperation situation and development trend of China-Africa renewable energy projects, in 2023, commissioned by the National Energy Administration, the China Renewable Energy Engineering Institute compiled statistics on the renewable energy cooperation projects carried out by 18 major Chinese energy enterprises on the African continent in the past ten years (2013 - 2023), and a total of 117 projects were collected, covering hydroelectricity, power grids, wind power, solar, geothermal energy, biomass and other types of renewable energy.

Among the 117 projects in the statistics, more than one-third of the projects have been signed (including projects under construction and successful bidding projects), with a total installed capacity of more than 10 GW, which are mainly concentrated in 7 countries, namely, South Africa, Ghana, Egypt, Côte d'Ivoire, Nigeria, Guinea and Zambia. The number of projects in the above seven countries accounted for nearly 60% of the total number of statistical projects. Currently, there are about 19 GW projects under active development for the preliminary stage.

## Typical Hydropower Project Cases

### Western Africa—The Inauguration Du Barrage Hydroelectrique De Soubrein in République de Côte d'Ivoire

❖ Located in the west of Cote d'Ivoire, the Inauguration Du Barrage Hydroelectrique De Soubre in Côte d'Ivoire has a total installed capacity of 275 MW and a total reservoir capacity of 83 million $m^3$, totaled investment of \$ 560 million. The project is an earth-rock fill dam with a maximum height of 20 m and a total length of 4. 5 km. The Chinese enterprise is responsible for the EPC work of the project, undertaking the hub project, transmission and substation project and other related work. In May 2017, the first unit of the power station was connected to the grid and the project was completed in November of the same year, with a total construction period of 56 months, which is 8 months ahead of the contracted construction period.

❖ The Inauguration Du Barrage Hydroelectrique De Soubre provides about 14% of electric energy to the whole country of Côte d'Ivoire. It effectively guarantees the stable development of the national electricity industry, improves the electric energy structure of Côte d'Ivoire, and advances the development of the renewable energy industry. During the construction process, the project has employed more than 5,000 local employees, with a localization rate of 85. 3%, and trained a large number of senior managers, construction technicians, unit installation and power plant operation and maintenance personnel, which has vigorously promoted local employment. The project has won the "China-Africa Cooperation and Development Award" issued by the Embassy of Côte d'Ivoire, celebrating a key project of the national energy balance strategy of Côte d'Ivoire. The government of Côte d'Ivoire evaluated the project as "a model of economic and trade cooperation between China and Côte d'Ivoire", and awarded the Chinese party in charge of the project the "National Medal of Honor" of the Republic of Côte d'Ivoire.

The President of Côte d'Ivoire attended the completion ceremony of the project

## Typical Hydropower Projects

### Eastern Africa—Kafue Gorge Lower Hydropower Station in Zambia

✣ Located in Kafue Basin, a tributary of the Zambezi River in Zambia, the Kafue Gorge Lower Hydropower Station has a total installed capacity of 750 MW. Started in November 2015, the first unit was connected to the grid in June 2021, and in March 2023, all five units were combined to the grid.

✣ The Kafue Gorge Lower Hydropower Station is the largest project developed by Zambia in 40 years, which is hailed as the "No. 1 Project" of China-Zambia cooperation, and is also an exemplary model project in the BRI. The Kafue Gorge Lower Hydropower Station is of paramount significance to the stability of power supply and sustainable development in Zambia. As of the end of November 2023, the cumulative power generation capacity of the hydropower station has exceeded 7. 928 billion kWh, which increases the power supply of Zambia by about 38% and reduces the carbon dioxide emission by about 7. 062 million tons. Chinese-fund companies have supported the development of local education, health, and infrastructure by running schools, drilling wells, and building roads and bridges. The implementation of the project has created about 15,000 local jobs and trained more than 160 technical workers. The permanent works of the project include the owner's operation village, permanent power generation works, etc. The boarding secondary school is the most core functional building in the operation village project.

The boarding secondary school was built in the supporting project

## Typical Hydropower Project Cases

### Southern Africa—Caculo Cabaça, Angola

❖ Caculo Cabaça is located on the Kwanza River, the mother river of Angola, and about 230 kilometers away from the capital city of Luanda. Equiped with an installed capacity of 2,172 MW, Caculo Cabaçais the largest hydropower station constructed by Chinese-funded enterprises in Angola at present, and also the greatest hydropower project under construction in Africa, which is hailed as "the Three Gorges Project in Africa". In May 2023, in the presence of Angolan President João Lourenço, Caculo Cabaça successfully completed the dam closure, which opened the prelude to the construction of the main project of the dam. In 2024, Caculo Cabaça has basically completed all the temporary construction works, and is carrying out the excavation of the dam and the plant, etc.

❖ Angola's power infrastructure remains to be improved, and a large number of cities, including the capital Luanda, have long suffered power shortages. After the completion of the Caculo Cabaça, the average annual power generation capacity will reach 8.566 billion kWh, which can satisfy more than 50% of Angola's domestic demand for power supply, lower greenhouse gas emissions by about 7.2 million tons per year, and reduce the consumption of non-renewable petroleum and coal resources by 2.733 million tons. The project can also significantly improve the conditions of water resource utilization with the functions of both peak shaving and flood control, which will contribute to the economic and social development of the local community. Since the construction of Caculo Cabaça, the project department has also carried out training for local employees, helping to improve the skill level of local employees. During the construction and maintenance of the hydropower station, the project department has created nearly 10,000 jobs for local people.

Project construction site

## Typical Wind Power Project Cases

### Southern Africa—De Aar Wind Power Project, the Republic of South Africa

❖ Situated near the Township in the Northern Cape Province of South Africa, the De Aar Wind Power Station Project is the earliest wind power project integrating investment, construction and operation of Chinese power generation enterprises in Africa. It has installed 163 sets of 1. 5 MW wind turbines independently produced by China, with a total installed capacity of 244. 5 MW and a cumulative investment of RMB 2. 5 billion. The project commenced in October 2015. The first wind turbine was successfully lifted in September 2016, the lifting and commissioning of 163 wind turbines was fully completed in August 2017, and the project was put into operation in November of the same year.

❖ The project delivers about 760 million kWh of renewable electricity to the South African power grid in a single year, and also provides more than 700 local jobs, creating new opportunities to improve local livelihoods and promote economic and social development. Chinese enterprises have actively integrated into the local community, by establishing four early childhood education centers in De Aar to provide an educational platform for children from poor families and investing R4. 5 million annually to subsidize poor university students to complete their studies. By the end of 2023, 112 poor university students with good grades have been funded by the scholarship program. A special community fund has also been set up and medical buses have been donated to provide free medical care for over 9,000 community members per year.

An early childhood education center established by the Chinese company

## Typical Solar Project Case

### Central Africa—Sakai PV Power Plant Project in Central African Republic

❖ Rested in Bangui, the capital of the Central African Republic, the Sakai PV power plant is the first PV power plant in the Central African Republic. During the Beijing Summit of the Forum on China-Africa Cooperation (FOCAC) in September 2018, China and the Central African Republic (CAR) built a consensus on a China-aided PV power plant project, which was marked with a leading bilateral cooperation project by the Central African government. Designed and constructed by Chinese enterprises, the project officially commenced in April 2021, with a total installed capacity of 15 MW, configured with a 5 MWh energy storage system. It is designed for a total annual power generation capacity of 76 million kWh, which was successfully combined to the grid in June 2022.

❖ Bangui primarily relies on diesel and hydroelectric power generation with higher diesel costs and much slower development of hydropower. The photovoltaic power station project construction is better-known for its short cycle, environmentally friendly and larger installed capacity, which can effectively tackle the problem of local power shortage in the first place. In the year when the project began to generate electricity, the power station met about 30% of the electricity demand of Bangui city. Further, the project also provided about 700 employment opportunities during the construction, trained a large amount of PV power plant construction personnel for the local community and helped local workers master multiple skills. "Nowadays, some people in the community have operated new kiosks and others have run new small restaurants, where people can visit as usual at night. The PV plant has livened the community up, and I'm confident it will be even better in the future." Girabe, a PV plant line installer who lives on the outskirts of Bangui, exclaimed.

Local employees participated in the project construction

## Typical Solar Project Case

### Northern Africa—Noor Midelt Solar Therma Plant Phase Ⅲ Project in Ouarzazate, Kingdom of Morocco

❖ Noor Midel Phase Ⅲ 150 MW solar thermal plant power station is located in Noor Midel Solar Power Park, Ouarzazate, in the eastern of Morocco. The power station is the large-capacity tower molten salt photothermal power plant project, boasting about 1,320,000 $m^2$ mirror area on the site, 243 m in the height of the heat-absorbing tower, 7. 5 hours of heat storage time. It is totally deployed 7,400 giant Heliostats and is designed to deliver 530 million kWh of renewable electricity per year. For the first time in the world, the project adopts a hybrid concrete and steel structure for the light tower and also combines a series of new technologies, techniques and materials developed by Chinese enterprises. More than 1 million local residents have been of benefit.

❖ The project was constructed using the EPC general contracting method, with Chinese companies responsible for design, construction and equipment supply. The project commenced in May 2015 and was completed in October 2018. The project has been awarded Morocco's "Five-star Quality Award", "Five-star Safety Award", "Social Contribution Award", "Economic Employment Promotion Award" and the like. During the construction of the power station, more than 60 local subcontractors and suppliers have been introduced, and hundreds of local enterprises have been involved in the surrounding area, creating nearly 14,000 jobs for the local community in total.

Cold and hot molten salt tank of power station

## Typical Biomass Project Cases

### Eastern Africa—Laibi Waste-to-Energy Project, Ethiopia

❖ Located in Addis Ababa, the capital of Ethiopia, Laibi waste-to-energy power plant has a total installed capacity of 50 MW, including 2 × 600 t/d municipal waste incinerator and auxiliary systems, 2 × 25 MW condensing turbine-generator units and auxiliary systems, and a 132 kV single-bus booster station. The project adopts the most global cutting-edge high-performance waste incineration technology and equipment. In this way, flue gas emission meets EU 2000 standards, disposes of 1,280 tons of waste per day in design with an annual waste treatment capacity of 437,500 tons and power generation capacity reaches 185 million kWh in a year. The project officially commenced in September 2014, entered the full commissioning stage in April 2017, and was completed in September 2017 for commercial operation.

❖ The Laibi waste-to-energy plant is known as the first cooperation project between China and Ethiopia in the field of environmental protection, and also the earliest waste-to-energy plant in Africa. After the completion of the project, on the one hand, it can tackle environmental pollution, on the other hand, it improves the level of local power accessibility. Laibi waste power plant has become a highlight of China-Ethiopia cooperation and a landmark project of China-Africa green cooperation. Ethiopian Power said, "This is a landmark project in Ethiopia and even in Africa". "The plant has a greater significance of treating garbage than power generation because in big cities like Addis Ababa, disposal of household garbage has become an increasingly critical livelihood issue, with extraordinary social significance."

Waste-to-Energy Plant Construction Site

## Typical Microgrid Project Case

### Eastern Africa—Somali Regional State Off-Grid PV Power Plant Project, Ethiopia

> "The four PV power plants undertaken by Chinese enterprises are the earliest projects to be completed and put into operation among the first 12 off-grid solar power plants in Ethiopia. It throws the first light on the villages where they are located. We would like to highly praise the perseverance of the Chinese constructors, which is the embodiment of the strength and responsibility of Chinese enterprises."
>
> —Sileshi Bekele, then Minister of Water and Energy of Ethiopia

- The off-grid photovoltaic power plant project in the Somali region is located near Collier village in the Somali region of Ethiopia, with a designed annual power generation capacity of 1.42 million kWh and a combined investment of about 14 million dollars. The project adopts an intelligent off-grid PV system, which can remotely monitor system operation data and adjust operation parameters. Based on a layered distributed design, the microgrid control system can realize millisecond-level rapid coordinated control, equipped with industry-leading energy storage converter and PV inverter, guaranteeing the ongoing and safe operation of the microgrid system. The key equipment used in the project, including the microgrid energy management system, coordination controllers, energy storage converters and PV inverters, are all proprietary Chinese products.
- The project has delivered benefits to more than 2,000 households in the neighborhood, enabling nearly 6,000 people to have access to renewable power energy. A local villager said, "Today is a memorable day, the darkness of night will be lit up by electric lights from now on, children can study in the electric lights, and we finally have access to the electricity we have been dreaming of". The World Bank said that the project built by Chinese enterprises has made a good demonstration of the "Light Up Africa" program, and the Bank will follow the example and continue to vigorously promote it in more than 200 other villages yet without access to electricity, and strive to realize the goal that 35% of the country's power supply will be powered by off-grid photovoltaic power generation in 2025, which will benefit 5.7 million households.

# 4

# Opportunities and Challenges of China-Africa Renewable Energy Cooperation

> Africa is a promising continent, and we are full of confidence in its prospects. The 21st century will certainly witness the common development and revitalization of China and Africa.
>
> —Wang Yi, Member of the Political Bureau of the Central Committee and Minister for Foreign Affairs of the People's Republic of China

> We value China's powerful support for Africa's integration, connectivity and FTA construction, and look forward to joining hands with China to promote the building of the China-Africa community of a shared future in the new era.
>
> —Moussa Faki Mahamat, Chairperson of the AU Commission

As the global demand for renewable energy increases, accelerating the development of renewable energy has become a global consensus. Africa showcases resilient economic growth and continues to unleash the energy demand. In the meanwhile, China has constantly made progress in industrial technology. As a consequence, the two sides have increased complementary advantage while Africa and China are seeking a higher level of cooperation, bringing in both sides new opportunities. China-Africa is carrying out renewable energy cooperation of its day. However, at the same time, Africa and China are also facing problems such as uncertainty in cooperation, funds shortage, and the urgent need to strengthen infrastructure and talent resources. Against this backdrop, China should provide mutual support and friendly cooperation with African countries and seize the new opportunities of renewable energy cooperation so as to jointly tackle the risks and challenges. Looking ahead, China and Africa will continue to deepen cooperation on renewable energy, consolidate and deepen the China-Africa strategic partnership through strategic coordination, mechanism construction, policy exchanges and project cooperation. In that way, we can promote the energy transformation and sustainable development of the two sides, and move forward hand in hand on the road to modernization, so as to build a high-level China-Africa community of a shard future.

## 4.1 Seize the Opportunities and Meet the Challenges Together

### 4.1.1 Opportunities for Cooperation

The International Community Has Built a Consensus on The Development of Renewable Energy Resources to Combat Climate Change

It is a general consensus of the international community in response to climate change by actively

developing renewable energy and promoting green and low-carbon transformation of the economy and society. At COP28 in 2022, 118 countries pledges to triple global renewable energy capacity and double energy efficiency by 2030 compared to 2022. UN research data shows that Africa's losses as a result of climate change were projected to be up to $ 440.5 billion during the period 2022 – 2030. In September 2023, the first Africa Climate Summit adopted the Nairobi Declaration. African Heads of State called on developing and developed countries to work together to reduce greenhouse gas emissions and urged developed countries to deliver on their commitments related to finance and technical assistance. It is recommended that the international community assist Africa in upgrading its renewable energy generation capacity from 56 GW in 2022 to 300 GW in 2030.

### Promoting a Community of a Shared Future and Helping Accelerate Africa's Transformation

China has always been participating in, contributing to and leading in the global green transformation. While promoting its own green transformation, it has proactively promoted international cooperation on climate and green development, actively shared its experience and technology in renewable energy development with other countries, and helped countries around the world to accelerate the development and utilization of renewable energy especially in developing countries such as Africa, so as to achieve the low-carbon transformation of energy. In September 2023, China announced at the Africa Climate Summit that it would develop and implement a South-South cooperation project to address climate change known as the "African Lights Belt". By leveraging the advantages of China's photovoltaic industry, China has created a Sino-African photovoltaic resource utilization cooperation demonstration belt, and help countries concerned in Africa to tackle the problem of difficult access to electricity so as to contribute to the green and low-carbon development.

### Africa Showcases Resilient Economic Growth and Continues to Unleash the Energy Demand

Africa has maintained a robust economy growth momentum, with the continent's average economic growth rate of 3.5% over the past two decades, second only to Asia. Despite facing multiple challenges such as COVID-19, geopolitical conflicts and global financial crunch, the African economy remains showing outright growth resilience and development vitality. With the accelerated pace of economic development in Africa and the gradual progress of urbanization, Africa will further increase the energy demand. The development of renewable energy is rooted in the priority development areas of the African Union. The African Union's Agenda 2063 proposes to take the enhancement of the response to climate change and the realization of sustainable development as one of the major objectives. As African countries remain committed to promoting the development of renewable energy, IRENA predicts that by 2030, Africa can meet nearly a quarter of its energy needs through the utilization of renewable energy.

### Africa Has an Abundance in Renewable Energy Resources and Great Potential for Cooperation

The African continent is vast and rich in renewable energy resources. The developable hydropower, wind

power and solar energy resources account for about 10%, 32% and 40% of the world respectively, but the existing installed capacity of hydropower, wind power and solar power accounts for less than 3% of the world. Its development level is relatively low in resource utilization, so there is a broad room for the development of renewable energy sources. In recent years, in order to meet the local energy and power security supply and economic development needs, Africa's major countries have proactively solicted support from the international community in renewable energy development funds, projects and other aspects. While China, faced with the U. S. and Western suppression and the intensification of international competition in the green game, actively promote the renewable energy industry going global, hence, China and Africa have great potential for cooperation.

### China Has Always Committed to Industrial Technology Development, Increasing Complementing Advantages

After years of progress, China has accumulated extensive technical experience in the wind power, photovoltaic and other renewable energy sector, complemented by a more complete industrial chain and production capacity layout, leading to certain cost and technical advantages. China has the world's leading manufacturing capacity of solar panels, electric vehicles and other new energy products, empowering it to become one of the major technology service providers in the global lithium batteries and solid-state batteries industry. With the continuous optimization of the China-Africa trade structure, the technological content of China's exports to Africa has improved significantly, with the exports of electromechanical products and high-tech products to Africa accounting for more than 50% of the total. Seizing the opportunity of South-South cooperation and the BRI, Africa and China have forged a partnership in the renewable energy sector, which can greatly leverage the strengths on both sides and achieve win-win cooperation.

### China Actively Promotes Technological Innovation to Enable Renewable Energy Continues to Reduce Costs and Boost Efficiency

China continues to promote technological progress in photovoltaic and wind power, such as the continuous acceleration of the process of localization and large-scale production of wind turbines, and the rapid improvement of the conversion efficiency of photovoltaic power. All have effectively lowered the average cost of a degree of electricity of the relevant projects, providing high-quality and low-cost renewable energy products and services for the popularization and application of renewable energy around the globe. In the course of the booming development of the renewable energy industry, China will further share reasonably priced wind power, photovoltaic power generation and other green energy products with Africa. And through a wider range of renewable energy project cooperation and co-construction, it can transform Africa's huge renewable energy resource potential into tangible economic growth, and improve the economy and accessibility of renewable energy in African countries.

#### China-Africa Relations Usher in The Most Promising Period Currently and it is a Great Time to Establish Cooperation in the Renewable Energy Sector

China-Africa relations are in the best period in history currently. Strengthening solidarity and cooperation with African countries is an important cornerstone of China's diplomatic policy and also China's long-term and firm strategic choice. Almost all African countries that have established diplomatic relations with China have signed the BRI cooperation documents with China, and Africa constitutes an important part of BRI. China supports the implementation of the AU's Agenda 2063 and its flagship projects, and actively participates in the implementation of the PIDA and other programs. Africa and China will jointly seek opportunities for cooperation in renewable energy, strengthen trade and investment facilitation, and create favorable conditions for in-depth cooperation in the renewable energy market, with a view to further boosting the quality and upgrading of China-Africa cooperation.

### 4. 1. 2 Challenges in Cooperation

#### The Uncertainty for Cooperation Has Been Intensified by Changes in the International Landscape

In recent years, the global energy industry has been ups and downs, as demonstrated by the formidable contradiction between supply and demand in the energy market, sluggish economic development and greater difficulties in financing, which have led to the rise in energy prices, supply shortage, and a steep increase in debt pressure in Africa, posing a certain pressure on China-Africa cooperation in renewable energy. It is an important task for the future China-Africa cooperation in the field of renewable energy that to participate in the development of renewable energy projects based on the principles of fairness and justice, mutual benefit and win-win outcomes, shy away from political and economic risks in the project development and construction, reducing social and environmental impacts, and help Africa establish a modern energy system with renewable energy-focused to facilitate green and sustainable development.

#### Renewable Energy Development Highlights the Problem of Financial Difficulties

Despite Africa has huge potential for renewable energy development, its further development and utilization are constrained by the financial crunch in investment. During the period 2000 – 2020, the African continent received the investment in renewable energy with an average of only 3 billion U. S. dollars in a single year. According to relevant research, the amount of renewable energy investment in Africa amounted to $ 12 billion in 2023 thanks to multiple efforts, which can double by 2022 and four times more than in 2021. However, according to CMP projections, Africa will need $ 72 billion per year before 2040 just for power generation, energy storage (including pumped storage and electrochemical storage) and cross-border transmission, but existing investments are insufficient to meet the demand for the potential of Africa's to-be-exploited resources and electricity.

#### It is Imperative to Strengthen Supporting Infrastructure and Talent Resources

The development of renewable energy is demanding supporting infrastructure in terms of grid flexibility,

technology R&D and innovation, logistics equipment, etc., and renewable energy projects require a large number of professionals in the design, construction, operation and maintenance periods. However, there are currently deficiencies in the local renewable energy technological infrastructure and human resources, and the supporting infrastructure constraints produce an impact on the stabilization of the renewable energy industry chain and the supply chain and the building of a renewable energy supply system. In addition, African energy industry practitioners have a low management level and technical ability, coupled with a lack of talent reserves, which can increase the project construction costs to a certain extent an drag on the speed of project progress and later maintenance.

#### Project Commercial Sustainability can be Improved

Most countries in Africa have small industrial and commercial volumes and low user tariff affordability. In order to promote the popularization of electricity, the government has been providing handsome subsidies for electricity users for a long-term run, leading to greater financial pressure, and there is a long-term inversion of the electricity sales price and feed-in tariffs in some countries. Moreover, there are more countries in Africa, the construction of the power market mechanism has deficiencies, the policy is not fixed, and Chinese enterprises lack understanding of the overseas commercial laws in African countries and fall short of tracking and response experience. All these factors bring greater risks in the recovery of electricity costs of investment projects, which constraints the Sino-African cooperation in electric power projects to a certain extent and weighs on the development of the power industry in Africa.

## 4.2 Charting a New Course of Cooperation Together

### 4.2.1 Cooperation Orientation

#### Hydropower is Helpful for Development and Hydropower Construction is Orderly Promoted

Africa is rich in hydropower resources, and hydropower, has great potential for development as a key development area of AU PIDA planning. As the country with the largest hydropower generation capacity in the world, China enjoys the advantage of whole industry chain cooperation, with its hydropower planning, design, construction, equipment manufacturing, operation and maintenance all reaching at the world advanced level. Hydropower has multi-dimensional comprehensive benefits such as promoting local socio-economic development in Africa, meeting the demand of the power market, improving infrastructure construction, upgrading the level of transportation, fostering the development of local building materials and construction industry, and improving the flexibility and reliability of the power system. Based on the principles of "green development, equality and mutual benefit, and win-win cooperation", China and Africa can explore cooperation in the development of hydropower bases and comprehensive utilization of renewable energy in river basins based on Africa's actual needs.

### Promoting the Transformation of Wind and Solar Energy and Upgrading Green Energy

Africa is the continent with the most abundant solar energy resources on Earth, and wind energy resources are abundant. China is the world's largest country in the development and utilization of wind power and photovoltaic. The country's combined installed capacity of grid-connected wind power and photovoltaic power generation hit 1.05 billion kilowatts in 2023, having created a completely new energy industry chain. The large-scale units and floating wind power in the wind power field have met internationally advanced standards and photovoltaic module production capacity has noticeable strengths in the field of solar. New energy power generation is characterized by small investment, fast construction and improving the people's well-being, etc. China-Africa renewable energy cooperation can be tailored to the local conditions with a combination of centralized and distributed development modes so as to embrace the demand for load development in different scenarios. China can assist African countries in carrying out the new energy resources census, development planning, project construction, consumptive research, etc., and actively explore the prospects for future development of offshore wind power in Africa.

### Strengthen Girds Structure and Enhance Resource Allocation Capacity

As Africa faces unevenly distributed energy resources and market demand, the power grid is an important and indispensable guarantee for facilitating the development of renewable energy and the optimal allocation of power resources in Africa. In order to achieve the optimal allocation of power resources in Africa, the AU is committed to building a highly efficient, economic and sustainable AfSEM and relying on the construction of power grid interconnection to build a regional backbone power grid in Africa. China's power grid has set the longest safety record of the world's mega-grid and has put in place and into service the world's strongest power transmission capacity and the largest scale of new energy grid integration. In the field of power grid cooperation, Africa and China carry out close cooperation in creating regional power grid interconnection, domestic power grid planning, construction and renovation and upgrading, providing a solid foundation for building a pan-African power market.

### Stabilizing System by Energy Storage and Diversifying Power Structure

As Africa is experiencing rapid development in the new energy sector, its priority is placed on bolstering the flexible adjustment capacity of the power grid. Energy storage will be one of the paramount measures to ensure the security and stability of the power system under the circumstance of large-scale and high-ratio development of new energy in Africa. China's energy storage technology is featured as developing in diversified ways and constant iterations, with the installed scale of pumped storage power plants ranking at the top in the world. Energy storage technologies such as lithium-ion batteries, compressed air energy storage, and supercapacitors meet internationally advanced standards. China and African countries can jointly carry out research on energy storage planning and configuration, explore and promote the diversified application and development of pumped energy storage and the novel energy storage in Africa. Both sides also need to

strengthen technical exchanges and cooperation in energy storage system integration, battery management system, and operation of energy storage power stations, etc., so as to help Africa enhance its independent construction capacity in the field of energy storage.

Benefiting People by Micro-grids and Improving Access to Energy and Electricity

By 2023, Africa has a population of exceeding 1.4 billion, leading to increasing demand for power consumption in people's daily life. The AU has taken improving power coverage, reducing the population without access to electricity and guaranteeing power accessibility as an important part in Africa's power development. China tackled comprehensively the problem of people without access to electricity in remote areas in 2015 by implementing the construction of power grid extension and renewable energy power supply projects in areas without the presence of electricity. Improving people's well-being constitutes an important part of China-Africa renewable energy cooperation. In solving the problem of power supply in powerless areas in Africa, by drawing on China's experience, a program based on Africa's actual needs was put forward to build a regional microgrid-type power system, which gradually tackles the problem of accessibility of power on the African continent by means of "program-demonstration-propagation".

### 4.2.2 Cooperation Recommendations

Deepening the Long-term Service of the Cooperation Mechanism and Playing the Role of Tracking, Coordination and Guarantee

We need to improve the long-term service mechanism, conduct better research on African regional and national top-level design policies, set up a better cooperation platform and absorb cooperation resources; assist in the formulation of reasonable and implementable renewable energy development plans according to the actual needs of African countries; join hands with the related African professional organizations to strengthen relevant policy research and share the experience of China and the third party in the development of renewable energy. We will help African countries do better in renewable energy resource census and development planning, and promote the development of renewable energy projects in line with Africa's infrastructure development planning, as well as Africa's key regions and countries' development strategies.

Facilitating Innovation in Green Financial Models, and Strengthening Precise Risk Assessment, Prevention and Control

We must increase green investment and financing support, encourage diversified innovative models, carry out medium-and long-term investment and financing activities for the renewable energy industry, and support livelihood projects that can enhance the accessibility of local electricity in Africa. On the premise of conforming to commercial principles, we provide financial support for Africa in renewable energy industry planning, project development, as well as operation and maintenance and other related areas. We encourage innovative green financial models, and based on the collection of project information, do better in evaluating and judging

the economic feasibility and financing risks of renewable energy green projects in Africa, so as to promote local green and sustainable development in Africa.

Building China-Africa Cooperation Demonstration Projects and Strengthening Technical Exchanges and Talent Cultivation

Through building renewable energy demonstration bases with African countries, we will strengthen the exchange, application and knowledge sharing of advanced technologies with African countries, and actively carry out diversified forms of international cooperation such as technical seminars, professional forums, joint research and project incubation. We need to explore the cultivation of local talents in Africa in various forms and help improve the renewable energy governance capacity, professional technology and labor skills of local practitioners in Africa through capacity-building training, joint office, setting up R&D, production and transportation bases, and introducing advanced equipment manufacturing lines and high-quality talents.

Jointly Strengthening Project Guarantee and Enhancing Project Economic Sustainability

China and Africa will work together to strengthen project security, sign corresponding investment protection agreements at the government level, shy away from the possible impact of political and social factors on the project, and mitigate the resistance of policy changes and other unforeseen circumstances to the implementation of the project. At the same time, relying on specific projects to carry out case studies, through the whole process of planning, design, construction and operation of the development of empirical evidence, we must make long-cycle tracking evaluation, in-depth analyze the development of renewable energy projects in Africa's actual construction and operational effect of the project and summarize the project's economic and social benefits so as to enhance the economic sustainability of the project.

### 4.2.3 Cooperation Vision

Seizing the Opportunities for Cooperation and Strengthening Strategic Coordination

At present, Africa and China's relationship is at the best in history currently. The two sides have gradually formed an all-round, multi-level and three-dimensional cooperation framework based on the BRI, the Forum on China-Africa Cooperation and other cooperation mechanisms. Based on the China-Africa Cooperation Vision 2035 jointly formulated by the two sides, Africa and China will promote the alignment of the BRI with the AU Agenda 2063 and the development strategies of African countries, continuously build consensus on cooperation, establish the direction of medium-and long-term cooperation in the field of renewable energy, and jointly push forward the formulation of a cooperation plan in the field of renewable energy, so as to promote the high-quality and sustainable development of China-Africa renewable energy cooperation.

Consolidating Mechanisms and Addressing Challenges Together

Africa and China will continue to deepen cooperation in the renewable energy sector under the framework of the BRI and the FOCAC. Focusing on the China-AU Energy Partnership, both sides consolidate the cooperation in

the renewable energy sector under the bilateral cooperation mechanisms of China-Egypt, China-Morocco, China-South Africa, China-Namibia and China-Algeria, and actively broaden the long-term cooperation mechanism with more African countries. These endeavors promote resource sharing, advantages complementarity, mutual benefit and win-win cooperation in the field of renewable energy. Moreover, these practices can safeguard investment and trade facilitation, strengthen the commercial projects' sustainability, and jointly address the challenges of cooperation, enabling China-Africa to maintain a steady and sustainable cooperation in renewable energy.

### Deepening Production Capacity Cooperation and Achieving Mutual Benefit and Win-win Cooperation

Since the inception of the BRI, Africa and China have accelerated their cooperation in production capacity and investment driven by relevant production capacity and investment mechanisms. The two sides will continue to uphold the principles of mutual respect, mutual benefit and win-win cooperation, strengthen mutually beneficial cooperation in infrastructure planning, design, construction, operation and maintenance. Both sides also need to integrate Africa's renewable energy development demands with China's industrial chain technological advantages, promote cooperation in research, project development, technology transfer and transformation of achievements. That way, China can support Africa to improve the technical standard system in the field of renewable energy, and enhance the quality of infrastructure capacity, as well as help Africa upgrade manufacturing industries and boost industrial competitiveness and create more jobs.

### Setting an Exemplary Model for Cooperation and Leading Green Development

Africa and China will focus on cooperation in the fields of hydropower, wind power, photovoltaic and other green energy to facilitate sustainable development on both sides. In light of Africa's renewable energy development needs, both sides shall fully leverage Africa's advantages in renewable energy resources, focus on the development conditions in an integrated manner, and build demonstration projects. These include large and medium-sized landmark projects, "small yet smart" high-quality projects, and comprehensive utilization projects combined with agriculture and mining, etc. Setting a cooperation exemplary model in exploring business models for sustainable development, can play a guidance role in providing reference and demonstration for China-Africa renewable energy cooperation, and help enhance the sustainable development capacity of African countries, making useful contributions to improving people's living standards.

### Shoring up International Cooperation and Promoting Common Prosperity

Africa and China have always been a community with a shared future. China and Africa have always shared weal and woe and stayed in the same boat united as one. China is the biggest developing country, and Africa is home to the largest number of developing countries. China and Africa are more in need of strengthening solidarity and cooperation than ever. China will continue to call on the international community to jointly focus on China-Africa renewable energy cooperation and development, fully highlight the spirit of cooperation in China-Africa friendship, adhere to the principle of sincerity, real results, amity and good faith and uphold the values

of the greater good and shared interests with the priority of greater good. These efforts are necessary for shoring up pragmatic cooperation in the field of renewable energy with African countries and building a closer China-Africa community of a shared future, thereby setting an exemplary model for the promotion of building a community with a shared future for mankind.

# Chronology of China-Africa Renewable Energy Cooperation ( during the period 2021 – 2023 )

## In 2021

**January 6** China and the Democratic Republic of Congo ( DRC ) signed a memorandum of understanding on jointly promoting the construction of the BRI, sending a positive signal to the outside world that China and the DRC are committed to seeking common development and prosperity and laying the foundation for mutually beneficial cooperation between the two countries in multiple fields including renewable energy.

**January 7** Botswana and China signed a memorandum of understanding on jointly promoting the construction of the BRI and other cooperation documents. The joint construction of the BRI will create new opportunities, broaden new fields and open up new prospects for the two countries to further consolidate the foundation of cooperation between the two countries in the renewable energy sector.

**Mid-August** The first batch of units of Kafue Gorge Lower Hydropower Station in Zambia constructed by Chinese enterprises were combined to the grid. With a total installed capacity of 750 MW, it is the largest single infrastructure project under construction in Zambia and the largest hydropower station in the country. The project has created new favorable conditions for local economic and social development and improvement of people's livelihoods, contributing to tens of thousands of local employment opportunities. The Sinohydro Group Training Institute built for the supporting project has opened 6 sessions locally, and more than 300 trainees have successfully graduated, becoming the backbone of this project and even the construction of various projects in Zambia.

**October 19** NEA and the AUC signed the *Memorandum of Understanding between the National Energy Administration of the People's Republic of China and the African Union on the China-AU Energy Partnership*. According to the MoU, the two sides agreed to establish the China-AU Energy Partnership and set up a joint working group. The two sides will carry out cooperation in the areas of policy and information exchange, capacity building, project cooperation and tripartite cooperation.

**November 22** Guinea-Bissau and China signed a Memorandum of Understanding on Jointly Promoting the Construction of the Silk Road Economic Belt and the 21$^{st}$ Century Maritime Silk Road, laying the foundation for

mutually beneficial cooperation between the two countries in various fields, including renewable energy.

**November 26** In order to introduce the achievements of China-Africa cooperation in the new era and look forward to the future prospects of China-Africa cooperation, the State Council of the PRC has issued a white paper entitled *China and Africa in the New Era: A Partnership of Equals*, which comprehensively summarizes the new concepts, practices and achievements of President Xi Jinping's diplomatic thoughts in Africa, interprets the all-around, wide-ranging and three-dimensional pattern of China-Africa cooperation with numerous figures and facts, and systematically sort out the fruitful achievements made by Africa and China in the field of renewable energy and other cooperation areas.

**November 29 – 30** The 8th FOCAC was convened in Dakar, Senegal, with the theme of "Sharing African Energy Development Opportunities, Promoting Pragmatic Cooperation for China-Africa Energy Projects". China and African countries held in-depth discussions on strengthening cooperation and seeking common development and built a series of overarching consensuss. The meeting adopted the Dakar Declaration and the Dakar Action Plan (2022 – 2024). The two outcome documents clearly put forward that China will strengthen pragmatic cooperation in the field of energy with Africa within the framework of the China-AU Energy Partnership, jointly improve the level of electrification in Africa, increase the proportion of renewable energy, and gradually solve the problem of energy accessibility, so as to promote the sustainable development of energy by both sides.

**December 13** The first meeting of the Coordination Mechanism for China-AU Cooperation in Jointly Constructing the Belt and Road was convened in order to implement the spirit of the 8th FOCAC and in accordance with the relevant arrangements of *the Cooperation Plan on Jointly Promoting the Silk Road Economic Belt and the 21st Century Maritime Silk Road between the Government of the People's Republic of China and the African Union*, the relevant departments of the Chinese side and the AU Commission exchanged views on cooperation in relevant fields, including energy, capacity investment and infrastructure, and further consolidate the consensus. The meeting signed *the MoU between the National Development and Reform Commission of the People's Republic of China and the African Union Commission on the Establishment of the Cooperation Work Coordination Mechanism for the Jointly Building the "Belt and Road Initiative"*.

## In 2022

**January 5** At the invitation of Eritrean Foreign Minister Osman, State Councilor and Foreign Minister Wang Yi paid a visit to Eritrea. During the visit, President Isaias and the Foreign Ministers of the two sides held talks and meetings, and both sides agreed to seize the opportunity of Eritrea's participation in the BRI to consolidate cooperation in energy, infrastructure and multiple fields.

**January 5** Morocco and China signed the "Cooperation Plan on Jointly Promoting the Construction of the BRI between the Government of PRC and the Government of the Kingdom of Morocco", the first country in North Africa to sign the cooperation plan of this kind with China, which further strengthens the foundation of cooperation between the two countries in the field of renewable energy.

**March** The first unit of the Zonggu Water Power Station, built by a Chinese company, has officially been connected to the power grid in Nigeria. The power station, which is an important project under the Belt and Road Initiative in Africa, has an installed capacity of 700 MW and can serve multiple functions, including power generation, flood control, irrigation, aquaculture, and shipping. The project has also created employment opportunities for over 4,000 locals, and plays an important role in improving Nigeria's power shortage situation, enhancing the stability and sustainable power supply capacity of the power grid, and improving people's living standards.

**June** The geothermal power plant in Sosian, Kenya, undertaken by the Chinese enterprise, was put into operation. According to Kenya's Vision 2030, the country would achieve 100% renewable energy power generation by 2030, of which the installed capacity of geothermal power generation is 1,600 MW, accounting for 60% of the country's power generation capacity. The Sosian geothermal power plant is of great demonstration significance.

**June 6** The NEA has commissioned the International Department of the China Renewable Energy Engineering Institute to take the lead in the preparation and operation of the China-AU Energy Partnership. The Institute is responsible for liaising with the relevant departments, energy enterprises and research institutes of the two sides, and carrying out capacity building, joint research and project incubation under the framework of the China-AU Energy Partnership, which effectively pushes forward the pragmatic cooperation in the field of energy.

**June 15** The Sakai PV plant, the first photovoltaic power plant in the Central African Republic, was connected to the grid. During the Beijing Summit of the Forum on China-Africa Cooperation (FOCAC), China and the Central African Republic built a consensus on the construction of the power station project. Undertaken by the Chinese enterprise under general contract, the power plant has an installed capacity of 15 MW, which effectively alleviates the difficulty of electricity consumption in Bangui and promotes the local socio-economic development.

**October 24** Under the guidance of the NEA and the AUC, CREEI organized the first session of capacity-building training under the partnership framework with the theme of "Large-scale Photovoltaic Power Plants and Photovoltaic Plus". More than 200 trainees from the African side participated in the program and this can help African trainees understand and grasp the application of relevant technologies in the field of renewable energy and enhance their management capacity.

**November 2–3** During the visit of the President of the United Republic of Tanzania to China, Tanzania and China signed a joint statement on the establishment of a comprehensive strategic cooperative partnership, laying the foundation for mutually beneficial cooperation between the two countries in multiple fields including renewable energy.

**Late November** COP27 "China Corner" venue successfully hosted a side event on the theme of "Renewable Energy Enabling Access to Electricity in Africa". At the event, all relevant organizations jointly

published the "Renewable Energy Promotes Electricity Access in Africa: Egypt Initiative" and the report of the study on Africa titled "Analysis of the Current Situation of Accessible Electricity in Africa and Case Studies on Off-grid Renewable Energy" under the framework of the China-African Union (AU) Energy Partnership.

## In 2023

**March – April** The NEA was invited to visit Namibia and Angola. During the visit to Namibia, the President of Namibia, Mr. Geingob, welcomed Chinese enterprises to actively leverage their own advantages and participate in energy development and project cooperation in Namibia; during the visit to Angola, China and Angola conducted in-depth exchanges on renewable energy development and pragmatic project cooperation, and hosted a symposium for energy enterprises in Angola to learn about the progress of cooperation on key projects.

**April 18** During the visit of the head of the Gabonese Republic to China, Cabon and China signed a joint statement on the establishment of a comprehensive strategic cooperative partnership, laying the foundation for mutually beneficial cooperation between the two countries in multiple fields, including renewable energy.

**May 26** The Democratic Republic of Congo and China signed a joint statement on the establishment of a comprehensive strategic cooperative partnership, laying the foundation for mutually beneficial cooperation between the two countries in various fields, including renewable energy.

**June 19** The Director of the NEA Mr. Zhang Jianhua met with Mr. Losienzo Ramohoopa, the Minister of Power of the South African Presidency, who was invited to visit the country. The two sides made an in-depth exchange of views on China-South Africa cooperation in electric power, renewable energy and nuclear power.

**June 20** The Director of the NEA Mr. Zhang Jianhua held bilateral talks with the visiting Januari-Yusuf Makamba, Minister of Energy, Tanzanian. The two sides discussed oil and gas, renewable energy, power cooperation and other issues.

**July 10** China and Guinea-Bissau signed a memorandum of understanding on jointly building the BRI. Both sides agreed to elevate their relations to a strategic partnership, laying the foundation for mutually beneficial cooperation in multiple fields including renewable energy.

**July 17 – 21** Mr. Abdelmagid Tebbon, President of the People's Democratic Republic of Algeria, visited China. The two sides agreed to consolidate their partnership and open up broader prospects for cooperation between the two countries in multiple fields by deepening cooperation in infrastructure, energy and other area. During the meeting, the energy authorities of the two countries signed a memorandum of understanding on renewable energy cooperation to consolidate exchanges in the field of renewable energy between the two countries.

**August 14** The handover certificate (TOC) of Zungeru Hydropower Plant Project, the Nigeria's largest hydropower plant, was officially signed, marking the completion and handover of the project to the owner.

**August 22 – 24** During the 15th BRICS Summit was held in South Africa, the leaders of the five countries

exchanged views on BRICS cooperation and major international issues of common concern and reached a broader consensus on the theme of "BRICS and Africa: Deepening Partnership, Promoting Mutual Growth, Pursuing Sustainable Development, and Strengthening Inclusive Multilateralism".

During the meeting, the governments of China and South Africa signed the "Letter of Intent on Deepening the BRI Cooperation between China and South Africa". Chinese enterprises and South African Power Corporation signed a strategic cooperation memorandum in response to a number of reform initiatives in the field of electricity in South Africa, and actively participate in South Africa's renewable energy project investment, help South Africa's power grid development and energy transformation so as to seek mutual benefit and win-win cooperation.

**August 22** Under the guidance of the NEA and the AUC, CREEI and AU Permanent Mission to China jointly organized the first project promotion conference of energy cooperation under the framework of China-AU energy partnership. With the theme of "Sharing Energy Development Opportunities in Africa, Promoting Pragmatic Cooperation in China-Africa Projects", the conference is aimed at promoting China-Africa energy project cooperation and encouraging Chinese enterprises to actively participate in the PIDA and the investment and construction of energy projects in Africa. The conference was attended by 43 envoys in China from more than 30 African countries.

**August 23** Under the guidance of the NEA and the Chinese Mission to the AU, CREEI organized a kick-off meeting for the construction of the China-Africa renewable energy project information database in Beijing. The conference was attended by the leaders and representatives from about 25 major financial institutions and energy enterprises, who held in-depth discussions on the objectives, contents and mechanism of the project database. Subsequently, the project information will be reported regularly every year, so as to keep abreast of the current situation and development trend of China-Africa energy cooperation and help potential cooperation projects to be incubated and implemented.

**August 31 – September 3** During the visit of the President of the Republic of Benin to China Patrice Athanase Guillaume Talon, Benin and China signed a joint statement on the establishment of a strategic partnership, which will further deepen the friendly and mutually beneficial cooperation in multiple fields and further consolidate the foundation of cooperation between the two countries in the field of renewable energy.

**September 4** The first Africa Climate Summit was held in Nairobi, Kenya. The representative of China announced in the speech that in order to implement the Declaration on China-Africa Cooperation on Combating Climate Change, China will develop and implement the South-South cooperation "Africa Light Belt" project to support African countries to combat climate change and green low-carbon development.

**October** Busanga Hydropower Plant, which was surveyed and designed by a Chinese enterprise, was put into operation. Tshisekedi, President of the Democratic Republic of Congo (DRC) attended the inauguration ceremony and cut the ribbon. The total installed capacity of the project is 240 MW. After all four units of the hydropower station are put into service, the average annual power generation is expected to reach 1.331

billion kWh, accounting for about one-tenth of the national power of the DRC. This will greatly propel the local economic and social development.

**December 8** COP28 was inaugurated in Dubai. During the conference, a seminar on the cooperation on the BRI green low carbon transition was held, focusing on China-Africa energy innovation cooperation. It was announced that the China-Africa Energy Innovation Cooperation Accelerator Project was officially launched. Under the framework of the China-AU Energy Partnership, the program will select and promote innovative cases and innovative technological solutions to help Africa's energy transition.

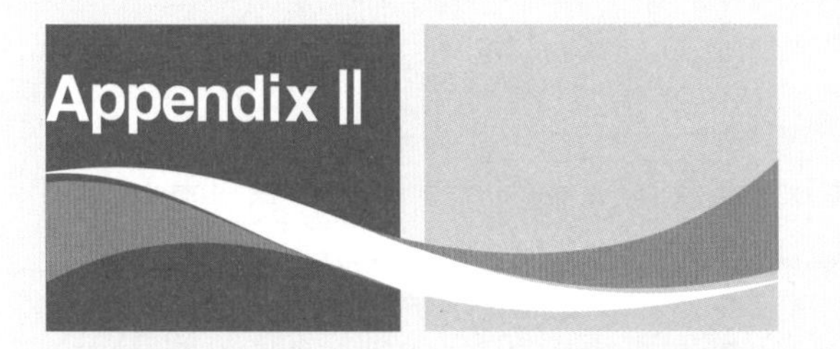


## Abbreviations

| | |
|---|---|
| APEC | Asia-Pacific Economic Cooperation |
| AfSEM | The African Single Electricity Market |
| AfCFTA | African Continental Free Trade Area |
| AFCONE | African Commission on Nuclear Energy |
| AFDB | African Development Bank |
| AU | African Union |
| AUC | African Union Commission |
| AUDA-NEPAD | African Union Development Agency |
| AFREC | African Energy Commission |
| AGHA | The Africa Green Hydrogen Alliance |
| BRI | the Belt and Road Initiative |
| BESS | Battery Energy Storage System |
| BRICS | Brazil, Rusia, India, China, and South Africa |
| CAPP | Central African Power Pool |
| COMELEC | Comité Maghrébin de L'Electricité |
| CMP | African Continental Power System Master Plan |
| COP28 | The 28th UNFCCC Conference of the Parties |
| CREEI | the China Renewable Energy Engineering Institute |
| DEI | Diversity, Equity and Inclusion |
| EAPP | East African Power Pool |
| EPC | Integrated Engineering-Procurement-Construction ( general engineering contracting model ) |
| EIB | European Investment Bank |
| FDI | Foreign Direct Investment |

| | |
|---|---|
| FOCAC | Forum on China-Africa Cooperation |
| G20 | Group of 20 |
| GDP | Gross Domestic Product |
| GWEC | Global Wind Energy Council |
| IEA | International Energy Agency |
| IRENA | International Renewable Energy Agency |
| IFC | International Finance Corporation |
| ICT | Information and Communication Technology |
| MoU | Memorandum of Understanding |
| NEA | National Energy Administration of the People's Republic of China |
| PIDA | Program for Infrastructure Development in Africa |
| SAPP | Southern African Power Pool |
| TOC | Taking-over Certificate |
| UAE | United Arab Emirates |
| UNDESA | United Nations Department of Economic and Social Affairs |
| UNESC | United Nations Economic and Social Council |
| UNCTAD | United Nations Conference on Trade and Development |
| UNECA | U. N. Economic Commission for Africa |
| WAPP | West African Power Pool |
| WBG | World Bank Group |

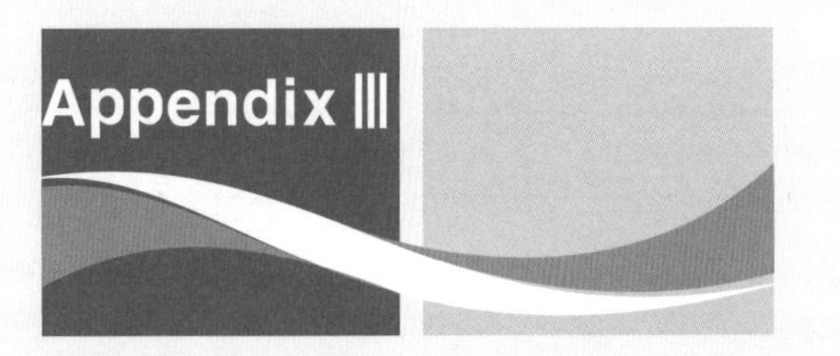


# Unit

| MW | Megawatt ( $1 \times 10^6$ W ) | TWh | Terawatt-hour ( $1 \times 10^{12}$ Wh ) |
|---|---|---|---|
| GW | Gigawatt ( $1 \times 10^9$ W ) | PWh | Petawatt-hour ( $1 \times 10^{15}$ Wh ) |
| kV | Kilovolt ( $1 \times 10^3$ V ) | EJ | Exajoule ( $1 \times 10^{18}$ J ) |
| kWh | Kilowatt-hour ( $1 \times 10^3$ Wh ) | GJ | Gigawatt ( $1 \times 10^9$ J ) |

# Statement



The relevant content, data, and viewpoints of this report are for reference only and do not constitute a basis for decision-making such as investment. China Renewable Energy Engineering Institute (CREEI) and African Union Development Agency (AUDA-NEPAD) shall not bear any responsibility for any losses caused by the use of the content of this report.

Unless otherwise specified, the statistical data of China in this report does not include the data of the Hong Kong Special Administrative Region, the Macao Special Administrative Region, and Taiwan Province. Due to rounding, there may be discrepancies between the total and the sum of the items.

A significant portion of the data in this report are derived from the United Nations Conference on Trade and Development (UNCTAD), the United Nations Economic Commission for Africa (UNECA), the United Nations Economic and Social Council (UNESC), the United Nations Department of Economic and Social Affairs (UNDESA), the International Renewable Energy Agency (IRENA), the International Energy Agency (IEA), the International Finance Corporation (IFC), the Global Wind Energy Council (GWEC), the Ministry of Foreign Affairs of the People's Republic of China, the National Energy Administration of the People's Republic of China (NEA), the African Union Commission (AUC), and the African Development Bank (AfDB), among others. Thanks are extended to all the above-mentioned organizations!